XA

Invasive Aliens

Invasive Aliens

The Plants and Animals From
Over There That Are Over Here

Dan Eatherley

WILLIAM
COLLINS

William Collins
An imprint of HarperCollins*Publishers*
1 London Bridge Street
London SE1 9GF

WilliamCollinsBooks.com

First published in Great Britain by William Collins in 2019

2020 2022 2021 2019
2 4 6 8 10 9 7 5 3 1

ISBN 978-0-00-826274-7 (hardback)

Typeset in Dante MT Std by Palimpsest Book Production Limited,
Falkirk, Stirlingshire
Printed and bound in Great Britain by CPI Group (UK) Ltd, Croydon CR0 4YY

To Georgia

'Turkeys, carps, hops, pickerel, and beer,
Came into England all in one year.'

Chronicle of the Kings of England
unto the Death of King James, Sir Richard Baker, 1643

Contents

Prologue

A Hornet's Nest

Tetbury, Gloucestershire, 2.30 pm Wednesday 28 September 2016
The wind had strengthened again and was blowing in short, powerful flurries. Graham paced up and down, his attention focused on the boundary hedges. Drawn by reports of sightings in the area, the team had last week netted several specimens of the invader that had been hawking for prey on ivy clinging to the trunk of one of the garden's cypresses. A reliable line of sight had been achieved and further samples dispatched for DNA analysis. But that had been all.

Graham approached the conifer once more. No activity here today. As he turned back though, something danced in his peripheral vision, something way up in the cypress closest to the house. Was it his imagination? He squinted for a better look. Sure enough, four, five, maybe six, large-ish insects were whirling about the highest branches.

What the heck were they doing? What on earth was inter-
esting them up there?

Not yet daring to hope, he reached for his binoculars.
Within moments Graham confirmed the target species and
shouted over to his colleague.

'Hey Gordon! This could be it!'

Confidence building, Graham scanned the canopy for a
few seconds, but saw only thick evergreen branches whipping
and twisting in the wind. Frustrating. He stepped back a
pace.

In that instant, a gust lifted the foliage enough to reveal
a tell-tale patch of light brown against the darker trunk. Had
he been standing a foot to the left or right, he would have
seen nothing. It was just a glimpse, but now he was certain
and made the call to the command centre: 'I think we've
found the nest.' All Graham could do now was wait. And
pray he was right.

The National Bee Unit – a division of APHA, the British
government's Animal and Plant Health Agency – has long
expected this unwelcome visitor from the far side of the
planet. Back home the Asian hornet is on the move, pushing
from its native range, on the border of northern India and
China, into Indonesia and South Korea. Soon Japan, too, will
succumb. The species was confirmed in Europe as early as
2004 when a nest was discovered close to Agen in southwest
France. No one really knows how it got there. The best guess
is that a year or two earlier some fertile queens had been
inadvertently shipped to Bordeaux in a container-load of
ceramic pottery from eastern China.

The hornet found the temperate European conditions,
similar to those back home, to its liking. Competition from

local hornets was minimal – the Asian variety is smaller than its native counterpart, more like a very large wasp; and far quieter. (Experts say the *chug-chug-chug* of an approaching European hornet calls to mind a lumbering Chinook helicopter. If so, the newcomer is a Stealth fighter.) And it relished the plentiful supply of food in the shape of other aerial insects. This is where the problem lies, for while the Asian hornet is partial to hoverflies, wasps and various types of wild bee, it really goes to town on the honeybee, swooping in on hives to pluck off its far smaller quarry. With a (compound) eye on efficiency, worker hornets behead their prey then dismember it, biting off the wings, legs and abdomen, before taking back to the nest only the thorax, stuffed with protein-rich flight muscles – perfect fodder for the developing hornet brood. It's the entomological equivalent of a shopping trip for prime rump steak, except the hornets do their own butchery.

In the worst cases, mobs of hornets will linger at the hive entrance, decapitating the emerging bees one after the other until they can move into the colony unchallenged, stripping it of honey, eggs and larvae. The bees back in Asia cobble together a defence of sorts, carpeting the hive entrance and swamping the intruders in a mass of shimmering abdomens. The heat-ball produced by the friction is enough to cook the intruder. The European bees have a go at this too but seem far less effective.

With the deck stacked in its favour, the Asian hornet has a field day. Over the summer, a single queen produces up to 6,000 young, almost ten times as many as typically found in a European hornet's nest. Most offspring develop into sterile worker females – the ones wreaking the damage to bee colonies – but around 900 turn into breeding males and 350 become reproductive females, known as gynes, which form the next generation of queens. While many gynes either fail to mate or perish over the winter, a few survive to start new nests themselves the following year. It needs but a handful of these foundresses to succeed for a population to take off, particularly if the new nests aren't detected and destroyed fast enough.

And that's what happened in France. When nests were discovered in built-up areas, local firefighters were called in to remove them, but otherwise the authorities ignored the hornet, perhaps reassured by its lack of obvious threat towards humans. The insect isn't especially aggressive and its sting no worse than that of the native hornet; in Europe perhaps a half-dozen fatalities to date can be attributed to Asian hornets – a statistic which compares favourably to deaths from wasp stings.

The honeybee population is less immune: in the Gironde département almost a third of colonies have been weakened or destroyed in a single year, one beekeeper losing 80 per cent of his hives. Dispersing at up to 80 kilometres each year – much further if the foundresses happen to hitch a ride in a passing vehicle – the Asian hornets have advanced unchecked over France. Along the way they have acquired a weakness for seafood, buzzing up and down coastal fish markets in search of small shrimps. It seems that any protein will do. By 2014, they had spread to Belgium, Portugal and Italy. One million of them were estimated to be in Germany. Then, in the early summer of 2016, they were spotted on the Channel Islands of Jersey and Alderney. It was only a matter of time before they reached the British mainland.

Changing agricultural practices going back a century have

devastated Britain's native bee and wasp populations, wiping out dozens of wild species and putting paid to many a honeybee colony. In recent decades, the threat to these important pollinators has worsened as crops have been drenched in pesticides and diseases have spread. Throw into the mix an exquisitely proficient bee-killer, and things have gone from bad to catastrophic.

For this reason, the UK government has adopted a zero-tolerance approach. The overriding objective of the Asian hornet response plan, the first of its kind in Europe, is the rapid interception and destruction of the insects before they get established. Beekeepers throughout the land are now on high alert, a select few tasked with active surveillance for the hornet, their colonies designated 'sentinel' apiaries. Meanwhile, members of the public are being encouraged to report, including via a mobile phone app, potential sightings of the visually distinctive invader. Unlike the native hornet, which sports a lively pattern of yellows and chestnuts on a dark-brown background, the Asian species is mostly black, save for a band of gold across the fourth segment of its abdomen. Added to that, it looks as if it's waded knee-high through a puddle of yellow paint. Until now, none of the thousand-plus sightings, routed to the National Bee Unit (located at Sand Hutton, just outside York) for confirmation, have proved positive. But vigilance is key and government-funded bee inspectors are on the front line. A contingency plan – regularly field-tested – is in place. But before it can be executed, a positive identification is needed. In other words, someone has to catch a specimen.

Tuesday 20 September 2016
'KILLER ASIAN HORNET INVASION'
'SWARMS OF VICIOUS HORNETS SET TO HEAD TO BRITAIN'
'MILITARY-STYLE OPERATION SPARKED TO DESTROY NESTS'
'BRITONS WARNED OF DEADLY ASIAN HORNETS THAT CAN
KILL FIFTY BEES A DAY'

The newspaper headlines that morning were typically understated. But they raised public awareness, and that was the
point. In a press release issued the night before, the government confirmed that an Asian hornet had been identified in
the Gloucestershire market town of Tetbury, and detailed the
urgent steps that would form its rapid response to the invader.
These included setting up a three-mile surveillance zone
around the town, establishing a command centre at an undis-
closed location and deploying bee inspectors to find the Asian
hornet nest and pest control teams to destroy it. Emphasising
the minimal threat to public safety posed by the hornets,
Nicola Spence – the government's top official for bee and
plant health – nevertheless promised 'swift and robust action'.

Monday 26 September 2016
The initial days of the operation proved frustrating. Each
morning the inspectors were briefed before heading off in
pairs to local apiaries armed with compasses, mobile phones
to record grid references, binoculars and purpose-built hornet
traps. At each site, the bee suit-clad inspectors hung around
for a couple of hours hoping to spot hornets. The team worked
off a list – provided by the Gloucestershire Beekeepers
Association – which included apiaries at Prince Charles's
Highgrove estate and the Westonbirt arboretum. Some sites
were located up to 20 kilometres from Tetbury, the initial

surveillance zone having been extended. Perhaps the first sighting had been just the tip of the iceberg. Perhaps *multiple* nests were out there. But it soon became apparent that the team was wasting time at distant apiaries – the real action was happening less than a kilometre from the original discovery. And not at beehives.

'I'm outside the Tesco's,' said one caller. 'I've just seen some of those wasp-things.' Sure enough, Asian hornets were confirmed buzzing about the red foliage of berberis hedging in the supermarket car park. More were clocked taking aphids from a willow coppice in a private garden nearby. And then this morning St Mary's Primary School, in the centre of Tetbury, reported them on ivy in the playground, the area in question rapidly cordoned off as the inspectors went to work.

In predation mode, the hornets seemed oblivious to the fuss. They were not afraid of humans. They just carried on hawking. Inspectors could walk right up to them and stick them in a net. Keeping an eye on the insects as they flew off was trickier. At first, in the hope of improving the hornets' visibility, the inspectors netted the insects, taped feathers to their abdomens and released them. But the hornets didn't cooperate; instead they stopped at nearby trees to remove the cumbersome tags and zoomed off. The inspectors got their eye in eventually and abandoned the approach. It was not worth the hassle, you could see a hornet flying without a feather and get a good feel for the direction. After much perseverance, lines of sight were collected. The team was closing in.

One o'clock. Wednesday 28 September 2016
Graham Royle, a Cheshire-based seasonal bee inspector, had arrived in Tetbury the night before. He was among those

providing relief to the 20-strong task force. More than 100 sites had been visited over the previous week, 50 traps deployed and dozens of fresh hornet sightings authenticated. A total of 94 separate observations would eventually be documented. Graham himself recorded a decent line of sight from a hedge that very morning, having trapped and released a hornet close to some ivy. Just before lunchtime, word came through from the command centre: 'We've got an intersect.'

As it turned out, the flight lines – four or five good ones – which were plotted on a map didn't converge neatly at one spot, but the criss-crossing did hint at a patch on the outskirts of Tetbury. Sixteen inspectors congregated in a car park. Teams of two, each provided with a map, started walking a section of this patch. It was not long afterwards that Graham, and fellow inspector Gordon Bull, found the nest. The garden with the cypress trees was a mere 600 metres from where the first hornet was spotted 23 days before.

Photos of the hornet's nest were emailed for confirmation to Dr Quentin Rome at the Muséum National d'Histoire Naturelle in Paris – he has studied the Asian hornet's advance with grim fascination for more than a decade. The nest was treated with a pesticide called Ficam B, an odourless white dust which is relatively harmless to humans but lethal to insects. Within an hour the nest was a mass grave.

The pumpkin-sized ball of chewed wood fibre was shipped to Sand Hutton for full analysis. The entire European population of Asian hornets belongs to just one of 13 subtypes known from the native range, implying a single introduction event, and genetic analysis later confirmed that the Tetbury hornets were of this same subtype. This suggested that they had indeed crossed the English Channel, rather than arriving independently from Asia. Further nests remained a possibility.

Residents were asked to remain vigilant and bee inspectors would stay on for a further two weeks until the Tetbury outbreak was declared over.

Woolacombe, North Devon. One year later
As he did every day at this time of the year, retired physics teacher Martyn Hocking headed up the valley to visit his bees. At his back the mid-afternoon sun was still high over the sea. It was best to visit the apiary while the occupants were out and about: there would be fewer to deal with. He heard the hum before he saw the hives; right now, they were drowning in late summer bracken and barely visible.

Giving the first hive a generous blast from his smoker, he lifted the lid and administered a dose of sugar syrup. A large dark insect flitted past. Too big for a honeybee drone. The thing hovered for a moment, offering an unmistakable view of a yellow-orange band on an otherwise black abdomen. One of Martyn's bees writhed in the grip of the larger insect, which seconds later darted off, vanishing into the emerald background as quickly as it had arrived.

I

Ecological Explosions

'Let it be remembered how powerful the influence of a single introduced tree or mammal has been shown to be.'

On the Origin of Species, Charles Darwin, 1859

The 16 July 1898 edition of the *Daily Mail* devotes a single paragraph to the revelation that unusual creatures were on the loose in one of London's better-heeled districts and defying all attempts at capture. According to the paper, 'The wild animals on Hampstead Heath have just received an unexpected addition in the shape of two monkeys which have escaped from custody and are now enjoying a free and open life on the salubrious heights.' A reward offered for the safe return of the simians – escaped pets from the nearby Bull and Bush tavern – proved unnecessary: a couple of days later the fugitives slunk back to the drinking house. Liberty hadn't agreed with them. 'They were in a deplorably dirty and woe-begone condition,' as one account had it.

It turns out that Hampstead Heath, a windswept expanse on a sandy ridge to the north of the city, is no stranger to the exotic. In 1944, monkey business was again reported from the Heath, the arboreal frolics of a pair dubbed Jack and Jill causing disturbance on this occasion. Things didn't end much better that time: Jill was shot and her dejected playmate handed himself in to the authorities. Then there was the young seal fished from a pond in 1926 having alarmed nocturnal anglers by 'barking like a dog with a sore throat'. Other tales tell of a phantom gorilla, of giant spiders, of marauding bands of wild pigs.

For the people of London, the Heath is a little piece of countryside on their doorstep, an oasis in which to de-stress and reconnect with nature. Keats, Shelley, Wordsworth and Constable were among the many poets and artists in thrall to its bucolic charms, its wild and unspoilt landscapes. So precious is this place that the threat of quarrying and house-building prompted an 1871 Act of Parliament protecting for ever 'the natural aspect and state of the Heath'.

But taking a stroll on the eastern end of the Heath one chilly autumnal morning in 2017, I was struck by just how much around me wasn't 'natural' in the sense of representing native British fauna and flora. The most obvious example was the resident flock of a hundred or so ring-necked parakeets with their frequent shrill calls. With eyes closed, I could have been roaming a Darjeeling tea estate. The birds originate from sub-Saharan Africa and South Asia and the stories to explain their introduction are every bit as colourful as their plumage. The most enduring legend is that they flew off the set of *The African Queen* during filming of the 1951 movie at Isleworth studios. Others point the finger at Jimi Hendrix for releasing a pair on Carnaby Street at the height of the Swinging Sixties. More probable is that London's parakeet population – the current estimate is 30,000 and growing – established itself after successive escapes from pet shops and aviaries. Perhaps not as numerous as in other parts of the capital, where parakeets are accused of beating woodpeckers and nuthatches to the choicest nesting sites, the Heath's contingent has been around for decades and seems to coexist happily with the locals. Then there were the grey squirrels. Victorians were the first to take a shine to the bushy-tailed rodents from North America and did their darnedest to spread them around the countryside. Woodland managers and red squirrel lovers alike have been gnashing their teeth ever since.

There's not much natural about the landscape either. Like every other part of Britain, Hampstead Heath has been managed and manipulated by people for centuries if not millennia. Cattle, sheep and goats – all first domesticated in the Middle East – have been raised here since the Neolithic period, suppressing forest regrowth and creating pasture. These days grazing duties fall to rabbits introduced from the Iberian Peninsula by the Normans. Or was it the Romans? Rabbits would give squirrels a run for their money for the title of 'world's worst pest' – just ask an Aussie farmer – but on the Heath they get a pass because their incessant munching helps preserve the acid grassland, and in turn a community of rare heathland organisms. From the eighteenth century, what little was left of the Heath's primeval woodland of pedunculate and sessile oaks, beech and birch – long exploited for timber and fuel – began to be 'enhanced' with Turkey oaks, horse-chestnuts, black locusts, rhododendrons, laurels and dozens more species from around the world. Early plantings were at the behest of aristocrats like Sir Thomas Maryon Wilson, Lord of Hampstead Manor, and the Mansfields of Kenwood House, whose estates constituted or bordered the Heath, but the 'parkifying', which included planting still more exotic trees and shrubs, continued long after the 1871 Act.

Meanwhile, the Heath's mineral-rich springs, dribbling out where porous Bagshot Sands meet impermeable London Clay, were exploited for drinking, laundry and their therapeutic properties. In time, the water, which had once collected in mosquito-infested swamps and bogs, was corralled into a series of ponds, which would later become a playground for bathers and anglers. Here, too, the roll-call of introductions is impressive from mandarin ducks and Canada geese to alpine newts and marsh frogs, from carp and catfish to red-eared terrapins. Some got here under their own steam. Most received a helping hand.

Human agency is suspected in particular for the arrival of two types of crayfish, among the Heath's more infamous aquatic denizens. The Turkish, or narrow-clawed, crayfish reached Britain in the 1930s, being joined in the 1990s by the red swamp crayfish from North America. Culinary motives are thought to have driven both introductions, with persons unknown considering Hampstead Heath the ideal place to rear them. They were spot on, for the two crayfish varieties multiplied, and are today well-entrenched in the ponds. This became painfully evident in 2012 when swimmers in the men's pond complained of being nipped on the toes and, according to one report, 'in altogether more sensitive places'. It gives 'bottom-feeder' a whole new meaning. Not long ago, the City of London Corporation, who manage the Heath, removed 500 red swamp crayfish from a single pond during routine maintenance over the course of just three days. With a single female producing up to 600 viable young in one go, total eradication is a tall order.

The point is that Hampstead Heath is populated by lots of living things from other parts of the world, many of them breeding and difficult to control. And I haven't yet mentioned some of the more notorious: plants such as Himalayan balsam, Japanese knotweed and giant hogweed; insect pests like oak processionary moth and harlequin ladybird; and virulent pathogens, such as Dutch elm disease, ash dieback and Massaria disease, which attacks plane trees. But, in this respect, Hampstead Heath is nothing special. I could have gone to pretty much any park in London, or indeed anywhere in Britain, and seen the same things, and far more besides. And it's not just parks: our rivers, lakes and streams; our forests and farmland; our estuaries and coastal waters; our homes and gardens; even our own bodies; all host a wealth of introduced species.

Many Brits pride themselves as stoic defenders of a green and pleasant land, boasting a record of resistance against aggressors dating back centuries, be it weathering the Spanish Armada or defying Hitler's Blitzkrieg. This patriotic fervour, and its clarion call 'to control borders', may in part explain the 2016 Brexit vote. Yet, a cursory examination of the natural world reveals that while many interlopers of the human variety have been kept at bay, our islands have throughout history been colonised by a succession of animals, plants, fungi and other organisms that apparently belong elsewhere. Indeed, it's often hard to sort out the native from foreign.

Philosophers and scientists have long noted the spread and impacts of introduced animals and plants around the world. In his *Naturalis Historia*, published around 78 CE, Pliny the

Elder wrote of Spanish rabbits, whose 'fertility is beyond counting', bringing such famine to the Balearic islands by ravaging the crops that the inhabitants begged the Emperor Augustus for military aid. Charles Darwin likewise observed the rampant spread of a European thistle across several South American islands during his nineteenth-century voyage on the *Beagle*. And, closer to home, in 1920 the Scottish writer James Ritchie highlighted the challenge to 'Nature's order' posed by 'many thoughtless introductions', arguing that 'the alien stow-aways which become established in a country include more economic pests than the native fauna they invade'. Yet, until the 1958 publication of *The Ecology of Invasions by Animals and Plants*, the march of non-native species was largely invisible to the wider public.

Its author, the pioneering British ecologist Charles Elton, believed that we faced a decisive battle whose outcome would determine the fate of the world. The book was an expansion of his series of BBC radio lectures entitled 'Balance and Barrier' and opened with a warning of the existential threat presented not only by nuclear bombs – the Cold War was by then gearing up – but by 'ecological explosions'. Elton defined these as 'the enormous increase in numbers of some kind of living organism – it may be an infectious virus like influenza, or a bacterium like bubonic plague, or a fungus like that of the potato disease, a green plant like the prickly pear, or an animal like the grey squirrel'. He identified 'the movement around the world by man of plants, especially those inten-tionally brought for crops or garden ornament or forestry' as among the primary reasons for the spread and establishment of new organisms. 'Just as trade followed the flag,' added Elton, 'so animals have followed the plants.'

In the decades since, international commerce has continued

to grow and the ecological explosions have kept on deto-
nating. In 2016 – the year Asian hornets were discovered in
Britain for the first time, sparking a military-style response
to make Elton proud – the first specimen of the Obama
flatworm slithered into Britain. Native to South America, the
species reaches seven centimetres in length and devours earthworms and other important soil invertebrates. This one
turned up in an Oxfordshire garden centre courtesy of a pot
plant from the Netherlands. The flatworm wasn't christened
in honour of the former American President but instead
derives its name from the Brazilian Tupi for 'leaf animal'.
That same year, three new infestations of the 'Asian super
ant' were discovered. Hailing from Turkey and Uzbekistan,
this social insect forms supercolonies numbering in the
millions and is also nicknamed the 'electricity ant' for its
inclination to congregate in power sockets and light switches,
chewing cables and threatening black-outs. Recent years have
also seen harlequin ladybirds from Asia kill off our home-
grown varieties; Pacific sea squirts carpet our marinas; and
buddleia and Japanese knotweed move down the rail network
more efficiently than most trains.

Public awareness of the issue is higher than ever before, with
sensational news headlines stoking our fears. Giant hogweed,
introduced as a horticultural curiosity from the Caucasus mountains in the 1820s, has been recast as Britain's 'most dangerous
plant' with sap that 'melts' a child's skin. 'Monster goldfish' are
on the prowl. 'Sex mad Spanish slugs' are terrorising our
gardens. Emotive terminology isn't just the preserve of tabloids:
even serious scientists will talk about 'demon shrimps' and
'killer algae' with a straight face. Some of the language has a
xenophobic flavour: introduced plants and animals are 'ex-pats'
or 'immigrants', which 'pollute' our pristine environment and

need to be 'bashed' and 'sent home'. Perhaps it's telling that the Nazis were among the first to take against non-natives, drafting a 'Reich Landscape Law' in 1941 banishing exotic plants from pure German landscapes. Some argue that the current fixation with non-indigenous wildlife is bound up with subliminal, and not so subliminal, antipathy to arrivals of the human kind. Concerns about non-natives and immigration to our small, overcrowded island are, they say, all of a piece. Even the term 'invasive species' has its drawbacks, perpetuating Elton's notion that we are somehow under assault, as if rhododendrons, grey squirrels and Asian hornets were working to a strict battle plan. The word 'alien', which remains in wide use, particularly among botanists, can have similarly unfortunate connotations. Worries about many non-natives can be whipped up unnecessarily, and sometimes for unsavoury political ends. But we shouldn't avoid talking about them: new organisms *are* arriving all the time, the pace of arrivals *is* rising and, yes, a handful of them *do* appear to cause problems.

There are other issues to consider. An invasive species is commonly thought of as a non-native organism whose population is increasing and spreading, and which causes, or may in the future cause, negative environmental, economic or social impacts. But what do we mean by 'non-native'? The usual understanding is that it's an organism introduced into a new country by people – on purpose or by accident – rather than getting there 'naturally' by walking, flying, swimming or wafting on the wind. In Britain, this often means anything brought here after rising sea levels cut us off from the European continent sometime between 7,000 and 9,500 years ago at the end of the last Ice Age. But what then do we call extinct fauna and flora that we have since reintroduced? The western capercaillie and European beaver, both around long before we became an island, were wiped out less than 300 years ago by hunting. According to the above interpretation, we couldn't treat them as 'non-native', yet both occur in twenty-first-century Britain thanks to human intervention. (The capercaillie was reintroduced to Scotland in the late 1830s, and successful releases of beaver have occurred at several locations across the UK over the last decade or so.)

At least with our current population of capercaillies and beavers we are sure how and why they're here. But, with many other species, it can be tricky to know when they first arrived, and whether people were involved. The sycamore, first recorded growing in the wild in Britain in 1632, is often regarded as introduced. Some say it was brought over in the fifteenth or sixteenth century, others suggest an earlier, possibly Roman, introduction, but either way we're looking at a non-native. Or are we? The problem is that sycamores, indigenous to central Europe, are fast-growing, fast-spreading trees well suited to Britain's temperate climate. While people have planted most of our sycamore stands, we can't exclude the possibility that sycamore seeds may have also taken root naturally from time to time having been blown across from the continent. If true, our sycamore population might comprise both natives and non-natives.

A question mark also hangs over the white-clawed crayfish. Considered our sole indigenous freshwater crayfish, and the focus of intensive conservation activity, some experts now suspect the crustacean was introduced for food in the thirteenth century. Even with things that we're 100 per cent sure are foreign, pinning how and when they arrived is a challenge.

Advances in DNA sequencing and analysis techniques are now shedding light on these mysteries. For instance, the presence on the Orkney Islands, off the north coast of Scotland, of a vole found nowhere else in Britain is something of a conundrum. The Orkney vole, as it's known, is an endemic subspecies of the common vole, a variety found on the European continent. So, did the Orkney vole scamper there naturally on a temporary land bridge from Europe before the last Ice Age, 20,000 years ago, and somehow manage to weather the chill while its British mainland relatives died out? Or, as seems more probable and has long been suspected, is the vole a far more recent introduction by people? Genetic studies seem to support the latter hypothesis, with Orkney voles shown to be more related to those in southwest France and Spain than to geographically closer populations. The voles are now thought to have arrived with humans on Orkney during the Neolithic period some 5,000 years ago in ships directly from the continent. The rodents may have stowed away in consignments of livestock fodder, or were intentionally brought as food items, pets or even for religious reasons.

Another difficulty with the notion of an invasive species as currently defined are those organisms, unarguably native, which behave in a manner that can only be described as, well, 'invasive'. Take hedgehogs. On the British mainland they couldn't be more popular; and faced by a catalogue of threats including habitat loss and agricultural pesticides, not to mention the risk of being flattened by motor vehicles, they're the focus of a nationwide preservation society. Yet on islands, Mrs Tiggy-Winkle goes berserk.

In 1974, someone on South Uist in Scotland's Outer Hebrides released half a dozen hedgehogs into their garden to keep down the slugs and snails. Within a couple of decades, the introduced mammal's population had swollen to 5,000 individuals, and had spread across causeways to the nearby islands of Benbecula and North Uist. The Hebridean hedgehogs eschewed garden pests in favour of the chicks and eggs of dunlins, ringed plovers, redshanks, lapwings and other shore-line birds, and are implicated in a 50 per cent decline in their numbers. Since 2001, nearly £3 million has been spent on removing the invaders. At first, they were culled with lethal injection, but tactics changed when animal rights groups kicked up a stink, so right now the hedgehogs are trapped alive and released on the mainland. In New Zealand, where introduced European hedgehogs have also run riot, the authorities have been less squeamish, eradicating them from the 86-hectare Quail Island.

Natives don't just act up on islands. Foxes and carrion crows, both indigenous to the UK, can obliterate nesting bird colonies. Red and roe deer can be every bit as destructive to woodlands and agricultural crops as their introduced counterparts: muntjac, fallow and sika deer. Common ragwort, bracken, brambles and nettles – natives all – often spread out of control, suppressing other plants, and are sometimes regarded as weeds. Even beech trees, indigenous to England, upset Scottish conservationists when their seedlings throw shade north of the border. A further complication is that a non-native might misbehave in one location but elsewhere – typically back home – cause no trouble at all, indeed may even be endangered. So, it's seldom fair to tar the entire species with the invasive brush. The hedgehog was one example but there are plenty more, such as the rhododendron, which spreads aggressively here but not in New Zealand, or the Japanese knotweed, whose penetrating roots are blamed for weakening buildings across Europe but which in Asia is surprisingly scarce.

Professor Helen Roy, a leading British expert on biological invasions based at the Centre for Ecology and Hydrology in

Oxfordshire, suggests that perhaps the only time when an understanding of whether a troublesome organism is native or not truly matters is when an incursion is still recent, or hasn't yet occurred, since it offers a possibility of intercepting it. Most of the time, though, the focus might be better spent on understanding, managing and, where practical, reducing any negative impacts of the troublemaker in question.

Despite these complexities, the term 'invasive species', as a label for fast-spreading, harmful non-native organisms, seems embedded in the common psyche. So how many do we have now in Britain? This is difficult to answer but thinking about it in terms of the invasion process can help. Scientists recognise several key hurdles any aspiring invader must clear. First, the organism needs an 'invasion pathway'; in other words it must get itself transported to a new region by humans. We may intentionally facilitate this, as with organisms brought for agriculture, hunting, horticulture, aquaculture, as biological control agents and for countless other reasons. But plenty of things use us to move around without our consent; think of plankton suspended in the ballast water of ocean-going vessels, the legions of wood-boring beetles holed up in internationally traded furniture, the seeds and spores peppering the mud of a tourist's hiking boots, the soil-borne invertebrates hitching a ride in plant pots. Crucially, the globetrotter must survive transit. This is no mean feat; the journey might take weeks during which the stowaway may be subjected to extremes of temperature, lack of food or moisture, and other privations. And that's even before the prospective invader has to contend with strict quarantine measures imposed by vigilant customs officials.

Before moving on, an important clarification is perhaps required. Should a new species reach our shores from its region of origin with no direct human intervention – be it by flying, rafting on an ocean current, blowing in the wind, catching a ride on a (non-human) animal, or in some other fashion – then this organism is classed as a 'natural colonist', and not counted in the statistics. A classic example would be the Eurasian collared dove, which started spreading west from Asia in the nineteenth century and was first recorded breeding in Norfolk in 1955. The distinction between natural colonist and human-mediated invader is not always as clear-cut. For instance, how to treat all those weird and wonderful new species set to colonise Britain – both on land and off our coasts – as the climate begins to warm? We may not directly be involved in their spread, but since human activities are driving global climate change, we are not innocent of this process. We might be happy to welcome recent natural colo-nists such as the tree bumblebee and small red-eyed damselfly, but are liable to baulk at accepting a malarial mosquito.

Importantly, though, an element of the natural about part of a new organism's journey movement won't earn it a pass as a natural colonist, if humans are known to have played a key role somewhere along the line. Take the Asian hornet and harlequin ladybird. Both insects are thought to have made incursions from the continent as a result of individuals blowing across the English Channel, but both only reached Europe in the first place as a result of human activities: the shipping of pottery from China, in the case of the hornet, and introduction as an agent of biological control, in the case of the ladybird.

On arrival, the introduced organism then has to escape and reproduce. Those that maintain a viable population, without further human intervention, are regarded as 'established' or 'naturalised', the rest dismissed as 'casuals'. (Incidentally, the term 'feral', according to Sir Christopher Lever, a British author of several well-known books on introduced animals, should be reserved for creatures that have 'lapsed into the wild from a domesticated condition', not simply escaped from captivity. Noting that populations of the American mink established in the British countryside are often referred to as 'feral', Lever insists that to regard the non-native mammal as 'domesticated' is 'preposterous and wrong'.)

Only around 10 per cent of organisms brought to a new country persist unaided in the wild. Disagreements over what constitutes non-native flora and fauna, along with the patchiness of data, have led to varying estimates of the number in Britain. The most recent figures, for 2017, suggest that 3,163 species were present in England, Scotland and Wales, of which 1,980 – mostly plants – had established and were reproducing in the wild. In Ireland, there are at least 1,266 non-native species, of which two-thirds are plants.

Finally, for something to be regarded as truly invasive, it needs to spread and expand its population enough to cause measurable negative impacts. On average, a minority of established non-natives register as a problem, although the likelihood of invasiveness varies: just 4 per cent of introduced insects in Britain are classed as invasive, compared with 32 per cent of non-native fish and 85 per cent of exotic plants.

This is all a long-winded way of saying that a tiny proportion of introduced species will ever earn the title 'invasive'. According to Helen Roy's team, which keeps a running score, in 2017 at least 275 (about 9 per cent) of the non-natives established in England, Scotland and Wales cause negative impacts. While around 5 per cent of the 1,266 introduced species recorded in Ireland are classed as invasive. These numbers will almost certainly climb.

So, let's turn to those impacts which can be classed as environmental, economic or social. Again, there is much debate, but the number one 'environmental' charge against invasives is that they harm natives, through predation, competition or, perhaps, by spreading disease. Invasive species are increasingly listed alongside habitat destruction, pollution and overhunting as key threats to wildlife. The 2005 Millennium Ecosystem Assessment says they're a major driver of biodiversity loss. A recent review of 247 kinds of plants and animals around the world that had vanished since 1500 found invasives to be the second most common cause of extinction (hunting, fishing or harvesting came first). For the amphibians, mammals and reptiles on this list of the disappeared, invasives were the number one culprit. Among the most frequent offenders were rats, cats and goats, along with diseases, such as avian malaria and chytridiomycosis, a fungal condition that is wiping out amphibians around the world.

Many of the most often cited cases of extinctions caused, or at least hastened, by introduced species come from islands. Famous examples include a near-flightless wren wiped out by the lighthouse-keeper's cat on New Zealand's Stephen Island; the dozen sorts of birds thought to have been extirpated by the brown tree snakes on Guam in the Pacific; or the eight varieties of endemic rodent dispatched on the Galapagos Islands by ship rats. A well-known non-island case comes from Lake Victoria in East Africa where the Nile perch was released by colonial Brits for sport-fishing in the late 1950s. This fast-growing predator has since been blamed for the loss of two-thirds of the lake's 300 types of endemic cichlid fish, although the introduction may merely have delivered the *coup de grâce* to dwindling populations already threatened by decades of over-harvesting and pollution.

Back here in Britain, concrete evidence for extinction is scarce, but we can't ignore two examples where introductions have threatened other species and could lead to their demise. While grey squirrels don't directly interfere with native red squirrels, they outcompete them for food, especially in deciduous woodland, and also pass on a lethal virus. The red's population crashed in the wake of the grey's arrival, so it's hard to avoid the conclusion that grey squirrels are a big part of the problem. Disease is also a reason that signal crayfish, brought from North America in the 1970s for aquaculture, are displacing Britain's white-clawed crayfish (which, as mentioned, may or may not be a true native). In this case, the signals pass on a fungal-like pathogen to the white-claws, which die within weeks of being infected. To be fair, 'crayfish plague' was already expunging the white-clawed population before signals came on the scene. Indeed, it was the signal's resistance to the plague that had recommended the crustacean to fish farmers in the first place.

There's no doubting, however, that even where an introduced species doesn't kill off a native, it can contribute to significant population declines. And if the impacts are only felt locally, it's still a concern. For instance, pirri-pirri burr, an Antipodean plant invader which reached Britain at the beginning of the twentieth century, probably as a hitch-hiker in sheep fleeces, will never trigger a national emergency, but in certain places – notably, Minsmere in Suffolk, and Lindisfarne island off the Northumberland coast – it threatens local wildlife.

Another problem, as some see it, wrought by invasive species is hybridisation. Even if non-natives seldom exterminate our home-grown wildlife, the tendency of many to interbreed with them is beyond the pale. Perhaps the best outcome is when the offspring prove sterile, although this represents, for the native, a waste of valuable breeding effort. More serious are cases where viable progeny arise and in turn back-cross with the indigenous species; this sort of thing can erode the gene pool, reducing the population's genetic variation and leaving it vulnerable to extinction. Hybridisation between native red deer and smaller introduced sika deer – an Asian variety – in parts of Scotland is a well-known example. Over time as the genes mix, red deer are starting to get smaller and sikas larger. As the two deer approach each other in size, this facilitates further hybridisation and risks accelerating negative impacts. Occasionally, hybridisation results in a more vigorous strain, as seems to be happening with the bluebell. Half the entire global population of this much-loved wildflower is found in this country, but a fertile hybrid has also established here, the result of a cross between the native bluebell and a Spanish variety introduced by horticulturists in the late nineteenth century. Many bluebells in Britain's gardens and urban areas turn out to be this hybrid, although even experts struggle to tell the difference and, for now at least, the hybrid bluebell does not seem to be invading woodlands. Indeed, recent research suggests that the Spanish bluebell is less fertile, and sets fewer seeds, than its British counterpart.

Ecologists also fear that invasive organisms could alter ecosystems in far more profound ways. These could include anything from changing water quality or soil nutrient levels to disrupting food webs, reducing pollination rates and generally messing about with the 'balance of nature'. Examples at random from around the world include the Mediterranean tamarisk tree, blamed for drying up marshes and salinising the soil in California, or zebra mussels altering nitrogen and phosphorus levels in freshwater habitats. One school of thought suggests that since ecosystems are dynamic and ever-changing, perhaps we shouldn't be too bothered. Such an attitude is simplistic and defeatist. Much of what humans – the most 'invasive' species of all – have done, from cutting down rainforests to spilling oil into the sea, from landfilling toxic waste to pumping out carbon dioxide, has upset ecosystems, and we need to understand and combat those negative effects however subtle. Right now – not for want of research – our understanding of how ecosystems function, how different organisms interact, and what makes these complex systems more resilient, or less, remains limited, with plenty of knowledge gaps left to fill. We are instinctively concerned each time a species is lost from a natural system through our actions (or negligence); we should also perhaps feel a similar disquiet whenever we cause a new one to be added.

If the invaders left people alone and restricted their impacts to the degrading of natural ecosystems, that would be bad enough – not least as we ultimately depend upon these systems for our survival and wellbeing. But some non-natives harm us directly. Notwithstanding the odd pinch on the privates from a crayfish, the obvious threat is their role as agents of disease. The most notorious example in history is offered by the Black Death, inflicted by a strain of bacteria originating in Asia which, from the fourteenth century onwards, has killed tens of millions across Europe, the Middle East and North Africa. While nothing on that scale has recently been visited upon us here in Britain, new parasites and pathogens are on the radar, many transmitted by mosquitoes and other biting insects. At the moment, it's a bit chilly for these to get a foothold here, but with climate change all bets are off.

Judging by the growing scientific literature devoted to the economic impacts of biological invaders, these species hurt our pockets too. Much of the cost arises from direct impacts such as insect pests reducing yields from agriculture and forestry, fish stocks wiped out by disease or the erosion caused when signal crayfish or Chinese mitten crabs tunnel into river banks. To the ledger we must add the eye-watering sums spent on preventing, monitoring and eradicating invasives. In excess of £5 million is spent every year in Britain removing Japanese knotweed alone. Various indirect impacts, trickier to calculate but just as real, and many times greater than the direct costs, can also be attributed to invasives. This is a complex area, but it boils down to the loss of valuable ecosystem services like nutrient cycling, pollination or flood prevention.

Overall costs incurred by invasive non-native organisms are estimated to amount to 5 per cent of the global economy. Across Europe, invasives inflict some £9-billion worth of damage every year. In the UK alone, the figure has been put at about £1.7 billion annually. Although these are ballpark estimates, resting on plenty of assumptions and subject to much debate, governments the world over are taking notice as never before. Invasive species are fast becoming public enemy number one. In 2016, the European Union banned 37 of the most problematic plants and animals from being kept or traded without a permit. These include signal crayfish, raccoons and American skunk cabbage. On this side of the Channel, the Great Britain Non-Native Species Secretariat was set up a decade ago and tasked with detecting and containing invaders, as well as helping to predict and prevent future incursions. Tackling troublesome non-natives is complex: the measures taken can be extraordinary and sometimes cause more problems than they solve, even hurting the very ecosystems they're intended to protect.

An emerging school of thought is suggesting that the threat of invasive species has been exaggerated, that we should stop worrying about non-natives and even welcome them for the benefits they can bring. At the other extreme, a growing band

of conservationists is going beyond simple calls for the eradication of non-natives to campaign for the deliberate
reintroduction of a menagerie of native British plants and
animals which have become extinct at the hands of humans.
To its critics, the 're-wilding' movement is pure eco-nostalgia.

For me though, most fascinating of all is that non-native
organisms, invasive or otherwise, from rabbits to rhododendrons, mink to muntjac, hold up a mirror to our own species.
Yes, the pace of invasion is higher than ever before but problematic non-natives aren't a modern phenomenon: they've
been with us from the outset, as unavoidable a corollary of
the human way of life as cleared forests and piles of garbage.
From the earliest settlement of our islands and first experiments with farming, through the Roman and medieval times,
and the age of exploration by Europeans, to the current period
of globalised free-for-all, the story of invasive species is the
story of our own past, present and future.

2

First Invaders

'But while men slept, his enemy came and sowed tares among the wheat, and went his way. But when the blade was sprung up, and brought forth fruit, then appeared the tares also.'

King James Bible, Matthew 13:25–26

For a million years a windswept peninsula in a corner of northwest Europe had seen various species of humans coming and going. The arrivals and departures were synchronised to the advance and retreat of continental glaciers, a dance choreographed by climatic change. They barely registered. A cluster of footprints here, a tidy pile of knapped flints there. Overwintering in caves, the people would emerge to gather shellfish from grass-fringed estuaries, pad through woodland in search of berries and nuts or pick at the carcasses left by lions and giant hyenas. The more ambitious, coveting the freshest meat, bone and fur, would rally family and friends in adrenaline-fuelled pursuits of deer, horse or mammoth.

Make no mistake, even the earliest people were unusual. Britain had never welcomed visitors quite like them, and over the aeons these experiments in humanity forged in the evolutionary crucible of an African valley generated ever more sophisticated results: the grunts of the most obtuse of cavemen took on deeper meanings; people fashioned better weapons and perfected their hunting techniques; they got the hang of butchery and learned to tame fire. Humans would turn their new-found skills on each other from time to time. Yet, for a great sweep of history, these pioneers – *Homo antecessor*, *Homo heidelbergensis* and maybe others – were but minor players on a stage dominated by rhinoceros and sabre-toothed cat, bison and bear. A low profile was often the best strategy given the monsters with which the land was shared. People were no more masters of their destiny than were grains of pollen in the air. And, every time the cold rushed back in and the fragrance of the dwindling forest was lost once more to the bitterness of endless tundra, so would humans again retire to more hospitable refuges in southern and southeastern Europe, abandoning the briefly colonised outpost to musk ox, wolverine and ice.

In the milder periods, when permafrost meltwaters inundated what would be known as the English Channel, the peninsula became an island. On one such occasion, some 125,000 years ago, humans found themselves shut out of the party altogether: things were warming up once again and a wealth of plant and animal species had spread back into Britain. But by the time people were on the scene, the land bridge from the continent had been claimed by the rising seas. Elephant, hyena, lion, deer, hippopotamus, elk and other animals had the place to themselves, enjoying a halcyon human-free interlude lasting 65,000 years.

Even *Homo sapiens*, the most successful hominid – in population terms, at least – to arrive in Britain was no great shakes at first. Originating perhaps more than 200,000 years ago, modern humans took their time getting here. Not for millennia would the most substantial exodus from Africa occur, with one wave of migrants moving along the Indian Ocean coastline towards southeast Asia, and eventually Australasia; another meandering north and west across the Middle East and Europe. When, from around 40,000 years ago, small bands of nomads, each with its own distinctive material culture, started to reach our shores, perhaps in seasonal visits, they found they'd been beaten to it.

Homo neanderthalensis had been eking out a living on Britain's cold treeless steppes for at least the previous 20,000 years hunting, with flint-tipped wooden spears, woolly mammoth, woolly rhinoceros and probably quite woolly bison. The Neanderthals, with their heavy eyebrow ridges, flared nostrils and stocky physiques, were well suited to the hostile conditions. Yet their days were numbered, the British contingent vanishing within a thousand years of the arrival of modern humans. No one knows why.

It's tempting to equate correlation with causation and accuse *Homo sapiens* of behaving as the archetypal invasive species, outcompeting and eradicating a vulnerable native with cunning and violence, perhaps passing on some disease for good measure. But the story appears more complicated: the significant amount of Neanderthal DNA in the modern human genome suggests a peaceful, even romantic, coexist-

ence between the two hominids across continental Europe
dating back 100,000 years. Could it be that the two varieties
of early human preferred to make love not war? In Britain,
at least, it seems they had little direct contact, and in any case
the argument is academic since dropping temperatures led to
the most recent glacial maximum about 22,000 years ago. Any
prospect of human existence was snuffed out for another ten
millennia. The ice crept forward, smothering everything,
wiping the sheet clean. Another fresh start.

The people who returned to a warming Britain from around
15,000 years ago could still be classed as hunter-gatherers, but
there was a greater sophistication about them, judging by the
plethora of artefacts and art left behind. They were dog-lovers
too, grey wolves having been domesticated to mutual advantage, possibly more than once, during or even before this most
recent Ice Age. Moving along the Atlantic coast, the humans
tracked herds of reindeer, horse, deer and elk north from
their southern European refugia. Some perhaps crossed the
English Channel in boats, while others may have sauntered
through Doggerland, an expanse of terrain today submerged
beneath the North Sea.

These people exploited natural resources with unprecedented intelligence: flint-tipped arrows of hazel, fired from
bows of elm, felled aurochs (wild ox), red deer and wild boar
with accuracy; the slipperiest of fish were trapped in river weirs
purpose-built from willow; birds and smaller mammals were
noosed and snared; a wider range of plants was collected,
stored and cooked than ever before. People were thinking ahead.
Fire was used to manage woodland. Freshly burnt clearings,
the ash festooned with appetising plant regrowth, could be used
to lure hungry game, which was much easier than tracking a
deer or boar through dense forest. Nevertheless, impacts on

the landscape were minimal. As in previous migrations people travelled light and, save for the plant seeds brought as food or stuck to clothing and bedding, few in the way of new species were conveyed to Britain during this period. Things though were about to change.

Danger. Tree felling in progress. A yellow warning sign greeted us as we approached the kissing gate. I had expected this: Hembury Hillfort's website requested visitors to 'observe cordoned off areas with red and white tapes', and please to 'not climb on timber stacks'. Thankfully, given my five-year-old daughter's enthusiasm for outdoors rampaging, neither woodpiles nor tape were in evidence today. The works programme, aimed at reducing root damage to the site's archaeology, was finished for the season. The tree clearance had a secondary function, to open up the view: that's what partly drew us here. Hembury didn't disappoint.

Twenty minutes later saw us picnicking amid bluebells at its southernmost tip. From the 240-metre-high bluff we were offered stupendous views across the Otter river valley towards the coast at Budleigh (the sea itself was lost in haze). The landscape was a hodgepodge of greens, interrupted here and there with the dull copper of a newly ploughed field, a yellow patch of oilseed rape, and the occasional pale minaret of wood smoke. Just visible to the west was Exeter, and beyond the grey eastern tors of Dartmoor from whose direction a brisk wind blew. Birds sang and robber flies buzzed. There was the faint drone of distant air traffic. Above us circled a pair of buzzards.

'What can you see?' I asked my daughter.

'Cows,' she replied, mouth stuffed with cheese-and-onion crisps.

Today's miscellany of embankments, trenches, mounds and other vestiges of Hembury's convoluted history confounds those wishing to understand it. The modern visitor is further disorientated by colossal beech trees which have erupted from the earthworks, clinging on with tentacular moss-covered roots. Yet its secrets are yielding to the archaeologist's trowel.

Hembury's strategic location and defensive qualities have long been recognised by those keen to defend themselves and command the region. It's a real Russian doll of a place: ostentatious double-ditched ramparts dug in the Iron Age, some 3,000 years ago, surround the entire three-hectare monument, which is perched at the edge of the Blackdown Hills in East Devon. Easy access to nearby iron ores and smelting works perhaps justified the investment in time and effort to shift the countless tonnes of earth by hand. Members of the Belgae tribe, from northern France and the Low Countries, subsequently laid claim to Hembury, making their own mark in about 50 BCE with additional defensive ditches and ridges across the centre of the fort. Then, in the middle of the first century CE, the Roman military too added Hembury to its network of forts – apparently taking it without a fight.

More fascinating still was Hembury's much earlier, Neolithic, incarnation, dating to around 6,000 years ago. This period was the focus of a pioneering series of digs in the early 1930s undertaken by the Devon Archaeological Exploration Society. The work was led by Dorothy M Liddell, a formidable and inspirational personality, and one of an emerging breed of female archaeologists. (A 17-year-old illustrator called Mary Nicol was one of Liddell's protégés at Hembury. Later, as Mary Douglas Leakey, she would make her own name with palaeontological discoveries in Africa.) Through meticulous excavations, Liddell detected signs of earlier inhabitation at

Hembury, including a causewayed (or interrupted) enclosure; post-holes denoting a once-grand timber gateway; the remnants of daub huts; shallow cooking pits, a metre and a half in diameter; and traces of a circular wooden building, possibly a guard house. Her team also recovered flint arrowheads and axes, and other stone implements, along with jet and greyish steatite beads and some of the earliest pieces of southern English pottery. Known as 'Hembury ware', the latter included simple round-bottomed bowls with lug handles, made using gabbroic clay, an orange-coloured mineral naturally occurring around the Lizard in Cornwall, 200 kilometres to the west. The finds hinted at a connection to an ancient and extensive commercial network stretching across the region and beyond.

But, for me, Liddell's most important discovery at Hembury were some charred grains of spelt, an ancient form of wheat. Carbon dated at roughly 5,000 years old, these represent some of the earliest archaeological evidence for the cereal anywhere in Britain. Liddell also turned up stone querns for grinding the crop into flour. Evidence of the importance of cereals in the diet of Hembury's Neolithic occupants was bolstered by the later discovery of 13 impressions of wheat grains embedded within some of the Neolithic ceramics.

How and why did a food plant native to the Middle East – 3,500 kilometres distant – come to be eaten atop a windy promontory in southwest England? The answer lies much further back in time.

Some 23,000 years ago, while Britain and the rest of northern Europe was gripped in an endless winter, people basking in the more benign climate of the eastern Mediterranean were gathering, grinding and cooking the grains of wild wheat, barley, oats and other grasses. It's possible that the most far-sighted and patient among them may have planted out some of their seeds and waited to harvest a crop. The evidence for such an innovation back then is patchy, but certainly by around 12,500 years ago farming communities had material-ised across the region.

The specifics of the transition from restless nomadism to a sedentary way of life based on cereal cultivation are still to be understood, but the shift is remarkably well documented in the Natufians, a people whose settlements are scattered across what is today Israel, Palestine, Jordan, northern Syria and southeastern Turkey. From about 14,500 years ago they started exploiting wild grasses such as emmer wheat and barley to make flatbread, beer and, later, animal feed. The transition from hunter-gatherer to settled farmer was by no means simple and direct. For some reason, the Natufians, having earlier taken up agriculture based on the intensive harvesting of wild grains, decided to resume a more mobile existence around 12,800 years ago. This about-turn has been linked to a colder period known as the Younger Dryas that reduced the natural availability of wild cereals in the Mediterranean region, forcing people to keep moving to fill their bellies.

Eventually, the Natufians and others returned to the culti-vation of cereals. By selecting varieties with the greatest yields, or those which thrived in diverse conditions, crops were grad-ually domesticated. Early agriculturalists benefited from a common mutation in wild wheat and barley that causes the grain-carrying spikelets to be more tightly gripped to the plant after ripening – just when they should be releasing them. In wild conditions, these 'non-shattering' mutants are at a competitive disadvantage compared to normal grasses which can spread their seed far and wide, but they lend themselves to being harvested and cultivated by humans. People learned to exploit other plants too, including flax, pea, chickpea, lentil and bitter vetch, intentionally planting, tending and harvesting them.

Scientists wonder whether the timing of this shift to crop domestication, which probably occurred independently in different places across the Fertile Crescent – as well as in parts of eastern Asia where wild varieties of millet and rice were the grains of choice – might not be a coincidence. One suggestion is that as the climate warmed at the end of the last Ice Age, and sea levels rose, so people were forced to higher ground where they would have encountered wild wheat and barley growing naturally. Levels of carbon dioxide were also increasing in the atmosphere – possibly due to its release from warming oceans – boosting worldwide plant production, including grasses such as cereals, and kick-starting what is often called the Neolithic revolution.

With the right kind of seeds, well-prepared soil and a favourable climate, the pioneer farmers soon found themselves amassing more food than they needed. This calorie boost, combined with a reduction in energy spent moving around, is thought to have ramped up human reproductive rates. A population boom led to civilisations across the Fertile Crescent, an 800-kilometre arc of territory encompassing the floodplains of the Nile, Tigris and Euphrates. But the discovery of farming may have set off a vicious cycle: the more people bred, the more food was needed and the harder everyone had to work. If they didn't want to, or couldn't, they might cheat or steal, requiring strong laws and even stronger rulers to keep the peace. Of course, there was nothing stopping rulers themselves from hoarding food and growing their own power in the process. At the same time, more and more of the landscape was turned over to crops which meant an acceleration in deforestation, erosion and other varieties of environmental degradation.

Unsurprisingly, given its peripheral location and challenging climate, Britain wasn't an early adopter of agriculture. By the time wheat and barley made their appearance here some 6,000 years ago – and those precious spelt grains were being hoarded in a primitive hut on Hembury hill – the world's first city of Uruk was already rising from the Mesopotamian floodplain. The farming of livestock also appeared in Britain, and the rest of northern Europe, around this time, again having been pioneered long before in the Middle East.

Goats and sheep are believed to have been domesticated from their wild ancestors – bezoar and mouflon, respectively – across southwest Asia from about 11,000 years ago. These low-maintenance creatures, compatible with a semi-nomadic lifestyle, were probably first kept for their flesh alone, and only later used for milk, wool and other secondary products. The fertilising properties of livestock manure was also noticed and exploited. Despite their benefits, sheep and goats would go on to become among the world's most destructive invaders, especially on islands where their relentless chomping wipes out rare plants and degrades ecosystems. Indeed, their unfussy diet, their rapid reproductive rate, their tolerance of a breadth of environmental conditions – the very traits which first drew us to them and of course to so many other problematic species – go a long way to explaining their world domination. At the last count, two billion sheep and goats roamed the planet.

Around the time that people first domesticated sheep and goats, cattle also joined the ranks of tamed ruminants. Cows were descended from the extinct wild ox, or aurochs. This was a spectacular beast, particularly the bull which stood nearly two metres high at the shoulder and sported fearsomely curved horns. Unlike the bezoar and mouflon, aurochs were already present in post-glacial Britain – indeed, they roamed the entire Eurasian landmass; however, domestication probably occurred in the Middle East. That's because early cattle were much smaller than our native aurochs, and DNA studies show that modern cows, including British ones, are genetically closer to Syrian aurochs than home-grown ones. In fact, today's entire global cattle herd – numbering some 1.5 billion cows – is believed to be descended from a founding stock of just 80 animals, likely to have originated in the Middle East. There's a good chance, however, that hybridisation would have occurred between local British aurochs and the smaller incoming cattle. Neolithic farmers may not have been thrilled about this: their petite cows, bred for milking, may have risked serious injury when attempting to birth an outsized hybrid calf.

Other modern domesticates with native British versions also seem to have derived from imported stock. These include the pig, whose ancestor, the wild boar, was widespread here before the advent of agriculture. Porkers are thought to have been first farmed in the eastern Anatolian region of modern-day Turkey about 10,000 years ago – along with a later independent domestication event in central China – and descendants of these Anatolian versions were subsequently brought to Britain. As with cattle, the amount of wild boar DNA in the genome of domestic pigs suggests frequent hybridisation between the two. To an extent, this may have benefited pig farmers, as crossbred versions may have been better suited to the more bracing local conditions in Britain, although too much of the 'wild' in a pig could make it a handful. A balance had to be struck.

Many of our supposedly native crops may also have come from elsewhere too. For instance, Britain's blackberries, raspberries, carrots and parsnips, as well as the perennial ryegrass, red clover and common vetch traditionally used as animal fodder, all probably derive from southern European strains.

Whether there's the whiff of the exotic about other domesticated species is less certain. For instance, the honeybee is thought to have originated in Asia, or maybe Africa, around 300,000 years ago, later spreading naturally across Europe, so the likely presence of this woodland insect in Britain before the most recent Ice Age would qualify it as native. Yet, the earliest known archaeological evidence for honeybee exploitation by humans in this country – as suggested by beeswax residues on seven pieces of Neolithic pottery found in southern England – dates to as recently as 4,000 years ago. That's several millennia after sweet-toothed pioneer farmers in Turkey, and later in central Europe, began gathering honey and wax from the insects, and possibly even domesticating them. So, we're left to wonder if Britain's first apiarists collected honey from wild bees or perhaps were using a tamer, introduced, variety that had been bred on the continent. In a sense, this discussion is somewhat academic, since pretty much all of our honeybees are today derived from southern European stock after parasitic mites devastated Britain's existing honeybee population in the early twentieth century.

So, how did the 'Neolithic package' – although this term for an apparent commonality of elements, including domesticated crops and livestock, along with other characteristic artefacts, is increasingly criticised as over-simplistic – reach our shores? Did Fertile Crescent farmers themselves migrate north and west, or was it just their agricultural practices that travelled, along with the wheat, barley, sheep, goats and other domesticated species upon which they were reliant? The question has been debated for well over a century, although recent research is beginning to support the former hypothesis. For instance, a genetic study published in 2018 found strong affinities between Mesolithic British and western European hunter-gatherers over a period spanning Britain's separation from the continent. The authors of this paper believe that British Neolithic people derived much of their ancestry from Anatolian farmers who followed the Mediterranean route of dispersal and entered Britain from northwestern mainland Europe. One thing is certain: when times were good, farming guaranteed a steady food supply and supported a burgeoning human population. In Britain, its practitioners rubbed along with nomadic hunter-gatherers for hundreds if not thousands of years, but the agricultural way of life, and the settled civilisation it supported, proved irresistible. So too, would the invasive species that profited from both.

The omens were there from the start. For millions of years, a spectrum of fast-growing, fast-spreading pioneer plants, both annuals and perennials, evolved to benefit from landscape impacts very similar to those that humans would one day cause. Many were adept at exploiting forest clearings opened up by fallen trees or recolonising habitats scraped clean by fires, glaciers, floods, landslips, volcanic activity and other natural disturbances. So, when the first farmers razed woodland and stripped soil bare in readiness for crops, they were teeing things up for a plethora of undesirable species. Commonly known as 'weeds', they have plagued us ever since.

Most troublesome of all were the weeds that resembled crops. These included darnel, a toxic grass which happened to be a dead ringer for wheat, and which infested the Middle East's earliest agricultural sites. Pastoralism only worsened the situation, as grazing and browsing livestock suppressed tree regrowth, maintaining the sort of open conditions favoured by weeds. What's more, just like crops, many weeds were adapted to thrive on the elevated levels of soil fertility resulting from all that extra animal dung.

British farmers, like their continental antecedents, set about annihilating the wildwood with their crops and livestock. Shifting agriculture was probably practised at first, with the felling of a few trees and controlled burning of understorey, followed by successive plantings of cereals. After a few seasons, the plot's soil nutrients were exhausted, forcing people to move on and repeat the destructive pattern. Anthropogenic deforestation was hardly a new thing – as we've seen, hunter-gatherers were keen on woodland openings – but its scale from the Neolithic onwards was unparalleled.

Trees were removed for reasons beyond the need for cropland: their timber was a source of both fuel and building material, while the clearances themselves may have held a symbolic value. Britain's vanishing woodland is reflected in changes in the incidence of particular pollen species in the archaeological record. As the representation of oak, elm, lime and ash dwindled, grasses, shrubs and wildflowers came to the fore. Invertebrate communities also changed, with a decline in specialist forest insects, including those associated with old or decaying timber, their place taken by varieties adapted to open and disturbed ground; dung beetles flourished thanks to livestock. Every so often a prolonged spell of climatic deterioration – as occurred between 5,000 and 3,500 years ago – would lead to a temporary abandonment of arable farming in Britain. Forests then had a chance to recover, although pastoral farming would still have been practised.

Of course, Britain's Neolithic farmers had their work cut

out dealing with the weeds that prospered in the denuded landscape. Many unwanted plants already lurked as seeds in our soil, just waiting for their moment in the sun; others were conveyed from further afield as contaminants of grain imports. The field, or corn, poppy, well-known to early Middle Eastern civilisations, is among the more familiar of the non-natives to have debuted in Britain around this time. The ancient Egyptians were taken by the striking blood-red blooms which infested their wheat and barley fields at harvest. The poppy's reappearance each year was a metaphor for rebirth and regeneration. The flower was woven into funerary bouquets and depicted on tombs.

Another arrival in Britain was charlock, or wild mustard, which was once described as the most troublesome annual weed of arable land. Indeed, an assortment of familiar crops including artichokes, flax, garden peas, leeks, lentils, lettuces and radishes may have started out as invaders of arable fields. Given that these are all fast-growing, short-lived species thriving on bare soil, their weedy heritage seems to fit. Even einkorn – one of the first types of wheat to be cultivated on a large scale – may have started life as a contaminant of emmer wheat crops. Furthermore, bread wheat, today's single most important variety, thanks to its easier threshing and greater grain yield, arose in the Fertile Crescent at least 8,500 years ago as a result of hybridisation between emmer and another weed, wild goat grass.

From a British perspective, some of the most important of the arable weeds were rye and wild-oats. Although originating in the Middle East, both seemed better adapted to our miserable climate and harsh soils, and often outperformed wheat and barley. So tenacious were these grassy invaders that by the Early Bronze Age, about 4,000 years ago, central and northern European farmers stopped bothering to weed them out and instead harvested them as crops in their own right. Domesticated varieties of both rye and oats were soon culti-vated for bread-making, for flavouring alcoholic drinks, and as animal feed. Wilder versions of the oat stuck around and remain intractable arable weeds to this day, in large part due to the similarities in appearance and lifecycle with those of crops. Selective herbicides are available but hand-weeding, or 'rogueing', of wild-oats is still practised on a small scale.

When Brits took to agriculture 6,000 years ago, the door wasn't just opened to invasive plants. Also waved through was an assortment of animal species adapted to living among people and exploiting their way of life. The house sparrow is a case in point. Remains of this small, gregarious bird have been identified in 10,000-year-old Natufian sites, suggesting sparrows long ago learned to nest in or close to buildings, purloining stored cereals and picking through the rubbish piles. By the Late Bronze Age, about 800 BCE, sparrows are known to have been present in central Sweden, so had probably reached Britain by then too. Today, they're one of the world's most cosmopolitan birds, outcompeting indigenous avians and proving a serious agricultural pest. In Russia alone, they've been accused of consuming a third of the annual grain produc-tion. During the 1950s the Chinese leader, Mao Zedong, even declared war on the sparrow, his scientists reckoning that, for every million birds killed, 60,000 extra people could be fed for a year. Chairman Mao's scheme backfired: the removal of sparrows resulted in plagues of locusts and other insect pests, whose populations the birds had helped suppress, which in turn led to famine. The Chinese government ended up reintroducing sparrows from the Soviet Union.

It seems therefore that house sparrows have a value in

agricultural systems and in Britain, at least, we're fond of them. The sparrow population has been falling of late: during the 1970s there were up to 12 million of them in the UK, but the population is now half that, with the worst declines in England. No one is really sure what's killing off sparrows. Possible factors include a reduced availability of invertebrate prey, a shortage of nesting sites and increased predation by squirrels, magpies and cats. In cities, high levels of nitrogen dioxide in the air, mainly from car exhausts, also seems to be a factor, with London alone seeing a 60 per cent decline between 1994 and 2004. All this has triggered urgent conservation efforts to save the sparrow.

Such measures won't be contemplated any time soon for the house mouse, another accomplished non-native invader, which originated up to a million years ago somewhere between the Middle East and northern India. The rodent was first drawn to the organic waste tips of hunter-gatherer settlements in the southern Levant at least 15,000 years ago and its population was primed to explode with the invention of agriculture. Recent evidence shows the house mouse sometimes shared the more mobile of the Natufian sites with a second species, the short-tailed mouse; however when people settled down for any length of time, the house mouse soon elbowed out its wilder cousin. By the Bronze Age, the rodent had scurried into western Europe but took a while to make its mark in Britain: the earliest records date from pre-Roman Iron Age settlements at Gussage All Saints in Dorset and Danebury Hillfort in Hampshire. The mouse seems to have got established after repeated introductions as a ship stowaway; by then Britain was well connected to the continent by the maritime trade and replete with granaries. Danebury alone boasted some 4,500 pits for storing crops, making it a house mouse heaven.

Along with rabbits, rats and grey squirrels, the house mouse shares the accolade of being among the few vertebrates to inflict both economic and social costs on a national scale. In addition to eating and fouling food stores, the rodent harbours a catalogue of unpalatable (and unpronounceable) diseases from tularaemia and typhus to leptospirosis and lymphocytic choriomeningitis. Humans have long waged a losing war against the species. These days baited traps and poisons tend to be used, but in times past barley cakes, spiked with black hellebore (a toxic variety of buttercup), would be placed at the entrance to their holes. Mice were also said to flee a censer of haematite stone and burning green tamarisk. But nature also provided a more elegant solution to the rodent problem.

The African wildcat's mouse-destroying prowess, along with its skill as a bird and fish catcher, may have been what recommended the species as the perfect household pet to the Egyptians more than 4,000 years ago. If true, that would make its tame version, the domestic cat, an early agent of biological control (the use of one organism to reduce populations of another). The sacred importance of cats in ancient Egypt is the stuff of legend with the feline deity Bastet worshipped as a goddess of fertility and the moon. The Greek historian Herodotus famously – but perhaps not altogether reliably – reported that the death of a cat prompted all those in the household to shave their eyebrows. The pet would then be embalmed. One cemetery unearthed at Beni Hasan in 1888 was said to contain the remains of 80,000 cats. A 20-tonne consignment of the corpses was later exported to Liverpool as fertiliser. One or two of the mummified moggies were saved for posterity by the city's museum. The human relationship with cats may predate ancient Egypt, with the suggestion that the felines began domesticating themselves during the Early Neolithic period; as sparrows and mice were drawn to Natufian grain stores and spoil heaps over 10,000 years ago, so cats were drawn to the sparrows and mice. A rise in the feline population may have been further sustained on proffered titbits from people, as well as rummaging through our mounting piles of rubbish.

Like the house mouse, the domestic cat first appeared in Britain towards the end of the Neolithic, with signs of the species at Gussage All Saints and Danebury Hillfort – just like those of its famous rodent quarry. Could it be that the cat's pest control qualities were appreciated in Iron Age Britain? Cats were, however, rare until medieval times. The earliest written record dates to the reign of the Welsh king Howell the Good (880–950 CE), who issued the edict that anyone slaying or stealing a cat was liable for a financial penalty calculated in terms of the equivalent cost in grain: 'The worth of a cat that is killed or stolen; its head is to be put downwards upon a clean even floor, with its tail lifted upwards, and thus suspended, whilst wheat is poured about it, until the tip of its tail be covered.' Today, an estimated nine million cats prowl Britain's towns and countryside, each year snaffling some 100 million prey items, including mammals, birds, reptiles and amphibians. One 1987 study from the village of Felmersham in Bedfordshire implicated cats in almost a third of house sparrow deaths. It seems old habits die hard.

Perhaps the greatest feline felony is a crime of passion. As with cows and pigs, keeping apart wild and domesticated versions often proves futile. The same seems true of pet pussies and Britain's own native wildcat, an endangered beast confined to the forested margins of Scottish moorland. The two versions have interbred so often that hybrids now dominate the wildcat population. Conservationists worry that too much domestic cat in the genome of the wildcat weakens it and leaves an animal which is already threatened by habitat loss and persecution close to extinction.

The arrival in Britain of a tabby of a different sort is also linked with the advent of Neolithic agriculture. Also known as the grease moth, the large tabby gets its name from the uncanny resemblance that its forewings bear to cat fur. With an appetite for dried dung, dead skin, old feathers, bits of straw and other unmentionable detritus, tabby larvae probably first hitched a ride here ensconced in livestock bedding. Suggesting that its natural habitat might once have been caves, the insect lurks in the gloomy recesses of stables and outhouses, where the larvae spin protective silken tubes about themselves then munch away undisturbed on their rarefied diet for up to two years before turning into adults.

A similar niche is exploited by dermestid beetles, many of whose 1,000 species and subspecies are spread by human migrations and globalised trade. Some are specialist scavengers on desiccated animal remains including hides, furs, feathers, tendons and bone, and a few are associated with Egyptian mummies, as well as with human remains from Middle Bronze Age sites in the southern Levant where the larvae drilled tunnels into the bone. Museum taxidermists still use these insects to nibble flesh from animal skeletons prior to display. Some dermestids could have reached Britain as early as the Neolithic period in the same way as the large tabby moth.

Among a number of non-native insect pests arriving in crop shipments is the grain weevil, a flightless species measuring around four millimetres when full-grown. Mated females each produce 150 eggs or more, which are deposited individually into grain kernels. The developing larvae feed there for up to six months before pupation, after which the adults chew their way out of the now-empty seed hulls. There's a theory that before agriculture came along the grain weevil's Asian ancestors lived on food scraps in bird or rodents' nests, before dispensing with wings altogether and becoming wholly dependent on human food stores. If true, this was a good move, as today the weevil plagues food stores worldwide, gorging on wheat, barley, rye, oats, corn, rice and millet, as well as a range of processed goodies from chocolate to pasta. The earliest western European record is from Early Neolithic Germany up to 7,000 years ago, and the insect is confirmed in Britain from the first century CE. Today, the UK alone spends an estimated £6.5 million annually on pesticides to control these and other non-native invertebrate pests of stored grains and fodder crops, including the saw-toothed grain beetle, foreign grain beetle and the red flour beetle, as well as mites and moths.

The unparalleled growth in human population and radical change in lifestyle unleashed by the Neolithic revolution benefited a different class of invading organisms; organisms that made their livelihoods not just among us, but on and even *inside* us. Harmful bacteria, viruses, protozoa, fungi, intestinal worms, ticks, lice and fleas, and myriad other nasties had always been present in the environment. For example, the bacteria responsible for tuberculosis, which still kills around three million people annually, was probably infecting the very earliest hominids in East Africa millions of years ago. The guts of hunter-gatherers are thought to have been crawling with roundworm, hookworm and other helminth worms, and their wounds quickly got infested with staphylococcal bacteria. In addition, a miscellany of animal-borne diseases may have infected humans before the Neolithic, from sleeping sickness and schistosomiasis to monkey malaria. But as soon as we started to form dense, semi-permanent, settlements, living side by side with livestock, and inadvertently drinking water contaminated by our own waste (never a good idea), harmful parasites and pathogens of all shapes and sizes were allowed to reach epidemic proportions for the first time.

For instance, the measles virus, in order to persist and spread, requires a sedentary population of up to half a million people with a continually replenishing supply of previously uninfected children. Malaria, yellow fever, diphtheria, leprosy, smallpox, influenza and the common cold are among a wide range of other 'civilisation diseases' thought to have benefited from our change of habits, many hopping from domesticated animal to human during, or after, the Neolithic. (The species-jumping may have gone both ways, with evidence that humans could have passed on harmful worms as well as certain other parasites and pathogens to their livestock, rather than vice-versa.) Furthermore, as we have seen, agriculture boosted populations of rodents, birds, invertebrates and other agents of disease. Even without close-living humans, grain stores, and herds of livestock, disturbance to the environment wrought by farming itself probably facilitated the spread of parasites and pathogens. For example, the deforested habitat resulting from slash-and-burn agriculture continues to favour malaria-carrying mosquitoes.

Britain's remote location, temperate conditions and relatively late adoption of modern farming may have helped its people avoid early epidemics. However, disease outbreaks probably became a fact of life by the Bronze Age with the increase in trade with the continent. Indeed, a catastrophic epidemic could explain the extraordinary results of a recent study on ancient human DNA across Europe which indicates that at least 90 per cent of the ancestry of Britons can be traced to the Beaker people. Named for their characteristic bell-shaped pots, this group originated in central and eastern Europe and arrived in Britain some 4,500 years ago, seemingly replacing almost the entire indigenous population. One suggestion is that the pre-Beaker Brits might have succumbed to a disease to which the Beakers were resistant.

Not everything that arrived towards the end of the Bronze Age and into the Iron Age was quite so unwelcome. By around 2,500 years ago, trade routes were beginning to extend to the Far East, courtesy of new imperial roads built by the Persians, facilitating a westward spread of previously unknown plants and animals. During this period, Brits may have got their first taste of a domestic apple, a species originating in the mountains of Central Asia, or ridden their first donkey, derived from wild asses in Egypt.

The woad plant, a member of the cabbage family prized as a source of indigo dye, was another Asian native appearing in Britain around this time. (Extracting the pigment was a complex process, involving huge quantities of leaves, a fair amount of an alkaline substance, such as lime – made by heating up chalk or limestone in a kiln – or stale urine, and a prolonged fermentation phase.) In *De Bello Gallico*, Julius Caesar's account of his seven-year campaign in the first century BCE to subdue the Gauls (another name for the Celts), he records that British warriors dyed themselves with woad to terrify their enemies. This was the inspiration for a blue-faced Mel Gibson in *Braveheart*. Like many of the best stories it has its doubters: the term Caesar used for 'woad' was *vitrum*, which also translates as 'glass', prompting some to suggest that Celts were in fact scarring or tattooing themselves.

Whatever the truth, pod fragments and seeds of woad have been discovered in the Late Iron Age site of Dragonby, near Scunthorpe in Lincolnshire, and it's believed the species was brought by Celts, via western and southern Europe.

The Romans may not have had a hand in bringing this particular plant to Britain, but that's more than can be said for a whole new wave of non-natives about to make their presence felt. Once again, momentous changes were afoot in this corner of northwestern Europe.

3

Romans and Normans

'This England never did, nor never shall,
Lie at the proud foot of a conqueror.'

The Life and Death of King John,
William Shakespeare, 1623

They didn't come for the weather, that was for sure. As Aulus
Plautius knew only too well, gales, incessant rain and a fleet-
destroying storm had scuppered Julius Caesar's attempts to
conquer the island in 55 and 54 BCE. But now, with orders from
the new and already beleaguered emperor Claudius ringing in
his ears, the general had no choice but to try again. So, when
the first Roman caliga squelched into British mud somewhere
along the southeast coast in 43 CE, there was a new determi-

nation to get the job done and, with 40,000 legionaries, auxiliaries and cavalry troops at his disposal, Plautius could hardly fail. Yes, some opposition would need to be dealt with. Caractacus, chieftain of the Catuvellauni people, was routed at the battle of Medway and his stronghold at Camulodunum – present-day Colchester – seized, but he fled to the west to fight a prolonged insurgency before his eventual capture. A few years later Boudica, the Iceni queen, also had a pop at the invaders, razing Camulodunum, along with Londinium (London) and Verulamium (St Albans). But she, too, succumbed. Rome would never conquer the entire island; however, within a century much had been brought to heel, with the Scots and other recalcitrants left to their own devices.

What Britannia lacked in climate and hospitable welcome was more than offset in mineral wealth: iron in Kent, silver in the Mendips and a generous seam of limestone from Oxfordshire to Lincolnshire, perfect for building roads and towns, aqueducts and bath-houses. Productive agricultural land was widespread too, although scant forest remained. Nevertheless, like all colonists, the Romans felt their new possession wasn't quite up to scratch.

The food in particular left much to be desired. Little in the way of fruit and veg was grown in Late Iron Age Britain. Notwithstanding the odd amphora of wine, olives, shellfish and other rarefied menu items that some pre-Roman elites are known to have imported, the locals had to content themselves with a diet heavy in oats and barley. A modest range of vegetables was cultivated, but dairy products were seasonal treats and meat a luxury. Most of today's familiar herbs and spices were absent. For the Romans, this just wouldn't do. Oats and barley were all very well for the subjugated – or as livestock fodder – but their own tastes were more refined.

The occupying power set about expanding the cuisine, introducing at least 50 new species of plant foods, most originating in the Mediterranean Basin. These included fruits such as peach, pear, fig, mulberry, sour cherry, plum, damson, date and pomegranate, along with almond, pine nut, sweet chestnut and walnut. Romans brought vegetables too, from cultivated leek and lettuce, to cucumber, rape and possibly turnip, along with new varieties of cabbage, carrot, parsnip and asparagus which already grew wild in Britain. Black pepper, coriander, dill, parsley, anise and black cumin added to a bonanza of outlandish flavours. Oil-rich seeds of sesame, hemp and black mustard were also among the arrivals.

Many introductions had supposed medicinal functions too. For the Roman historian, Cato the Elder, the cabbage surpassed all vegetables in that respect. Writing in about 160 BCE, he noted that it 'promotes digestion marvellously and is an excellent laxative'. Moreover, he insisted, there was nothing better than a warm splash of urine collected from a habitual cabbage-eater to treat headaches, poor eyesight, diseased private parts and sickly newborns. Another plant introduced to Britain for its therapeutic properties was Alexanders – the 'parsley of Alexandria' – a chunky lime-green relative of celery, which grew to 150 centimetres in height and was prized as aromatic vegetable and versatile tonic alike. The Romans may have been on to something here: recent chemical analysis of Alexanders reveals high concentrations of the anticancer compound isofuranodiene.

How many of these species were grown in Britain during the occupation rather than imported as ready-to-eat crops is unclear. The sweet chestnut, for instance, a staple of many a legionary's mess-tin, is absent from the medieval pollen record, suggesting it was grown here only much later. A period of hotter summers across northern Europe, including Britain, during the early years of Roman occupation may have favoured the growth of warmth-loving figs, mulberries, grapes, olives, pine nuts and lentils, albeit on a modest scale, perhaps in garden pots. By the time the Romans left, several introductions, including walnut, carrot and cherry, are known to have fully established themselves.

The origins of certain plants can be traced to Britain's first formal gardens, laid out during the Roman period. The best-known example is Fishbourne Palace in West Sussex, built in about 75 CE, whose outdoor space boasted tree-shaded colonnades and ornamental water features, along with geometric beds, fertilised with manure and bordered by a decorative hedging box. Fishbourne is now believed to have been the residence of a loyal Brit: Tiberius Claudius Cogidubnus, chieftain of the Regni tribe; if true, it was a handsome reward indeed for his allegiance to the occupying power.

A minority of Roman plant introductions are today regarded as invasive. One of them is probably ground-elder. This iron-rich perennial was cultivated both as culinary herb and for treating arthritis (another name for it is 'gout weed'), but once its spaghetti-like rhizomes got a foothold, ground-elder was near unstoppable. (Rhizomes are specialised subterranean stem sections capable of putting out both roots and new shoots.) To this day, up to £1 million is spent every year eradicating it from gardens. Some experts say ground-elder is native, but because the weed is usually found close to human habitation its presence here is generally blamed on the Romans.

As we've seen, sheep, cattle, pigs and goats were established in Britain prior to 43 CE, but the chicken – today the world's commonest and most widespread livestock species – was still

a rarity in this country, judging from its absence in the archaeological record. This may have been an artefact of the poor
preservation of their brittle bones and difficulties in identification. The earliest remains appear in Early Iron Age burial
sites (around 800 BCE), in Hertfordshire and Hampshire, and
their very scarcity may have perhaps been reason enough to
entomb these exotic birds from the Orient with the lately
departed. But when, where and why were people first drawn
to the red junglefowl, the chicken's probable wild predecessor?
No one knows for sure, but domestication seems to have
occurred somewhere in south or southeast Asia around 4,000
years ago, with tame fowl brought to the Mediterranean by
the eighth century BCE, reaching central Europe a hundred
years later.

Chickens and their eggs have always been eaten, but for
much of human history they've been as prized for their pugi-
listic prowess as for their gastronomic qualities. Cockerels, it
turned out, need scant encouragement to set at each other
with beak, claw and, in the older birds, wickedly sharp leg
spurs. The skirmishes have excited the bloodlust of onlookers
for generations. Cockfighting spread west across India and
the Middle East, the sport in turn captivating the Persians,
Greeks and the Romans. Chickens held a religious significance
too, the males symbolising the sun god in the Roman cult of
Mithras. Caged fowl would be taken on military campaigns
and their eating habitats studied for purposes of divination;
if your sacred chicken, when offered food, guzzled it down,
all augured well for the impending battle. Fowl-keeping in
Britain grew in popularity up to and throughout the Roman
invasion, albeit the preserve of a privileged few. Here, as
elsewhere, chickens were multifunctional, a source of food,
entertainment and devotion. Their bones are associated with

Roman temples, such as one at Uley in Gloucestershire dedicated to Mercury, and they regularly turn up in Romano-British graves.

Various other animals were imported for nutrition, status and religious reasons, with the remains of pheasant, peafowl, guinea fowl and donkey all found occurring in Roman sites. Elephants were the most impressive creatures brought to Britain; the Emperor Claudius used them to intimidate his new subjects soon after his victory – their stink had the added benefit of panicking enemy horses – although the tuskers' visit seems to have been fleeting. Archaeologists are intrigued by the discovery at Fishbourne and on the Isle of Thanet, Kent, of numerous bones of fallow deer, a variety hailing from the Anatolia region of modern-day Turkey. Analyses of the deer teeth at both sites indicate well-established, breeding populations, a finding that hints at the existence of what might turn out to be Britain's earliest deer parks.

As with so many non-natives, the story of fallow deer is far from straightforward since they vanished with the Romans around 400 CE. It was long assumed that the species only returned to Britain with the Normans, but recent radiocarbon dating work suggests they were around just *before* the Battle of Hastings. Either a few of the Roman deer hung on in the wild, or more likely, small-scale reintroductions, perhaps as novelty items, continued to occur over the course of succeeding centuries.

Sometimes creatures were kept for company alone. That seems to be true both for natives, such as ravens and crows, which were popular pets among the soldiers in Iron Age and Roman Britain, and for the more exotic. Examples of the latter included the Barbary macaque, a monkey whose bones have been recovered from Roman sites at Wroxeter, Dunstable and Catterick.

The Romans weren't averse to the odd invertebrate too, notably snails, new species of which were introduced as a delicacy. The pot lid, or Burgundy snail remains the most popular of several edible types that now support a multi-million-pound global *escargot* market. These days snails are largely absent from menus this side of the Channel, where they are regarded as vermin. Indeed, the 5,000 tonnes of molluscicide applied every year to keep them at bay could fill two Olympic swimming pools.

Most creepy-crawlies arriving and spreading during Roman times came unnoticed as hitch-hikers, such as grain weevils. The earliest British remains of these and other insect pests of food stores show up at sites in London and York dating to within the first decades of the Roman occupation, suggesting that infested grain was imported from Europe soon after the invasion. Invertebrate parasites of livestock and people flourished as new forts, towns and cities sprang up, and human population density grew. The Romans were known for their close attention to personal hygiene, with flushable latrines and heated bathwater. Yet, these measures failed to arrest the proliferation of tapeworm, liver flukes, roundworm and whipworm, along with swarms of fleas, lice and the odd bed bug. The widespread prominence of fish tapeworm, a gut parasite attaining nine metres in length, is something of a puzzle since the species is rarely evidenced in earlier, Bronze and Iron Age sites. Here, the Roman weakness for a peculiar condiment called garum may have been the cause. This fermented sauce, a blend of raw freshwater fish and herbs, left to rot in the sun, was traded across the empire and could have helped spread fish tapeworms.

From the late fourth century, the Roman Empire began to wither. Soldiers stationed in Britain were recalled to fight insurgencies on other fronts and by 410 CE the northern outpost had been abandoned. What happened over the next six centuries, traditionally dismissed as the Dark Ages for the paucity of written records, is vague. Roads and other imperial infrastructure disintegrated, vibrant towns and cities decayed, and trade declined, all slowing the influx and spread of new species. Yet, this was a period of great human churn as populations from Ireland, Scotland and other outlying regions of the British Isles moved into undefended territory, joined by continental immigrants, particularly from Scandinavia, the Netherlands and Germany. These movements of Angles, Saxons, Jutes and other peoples would have instigated fresh introductions, deliberate and accidental, but for now the details are lost in time.

The elite are always keen to improve upon what nature has provided and, when it comes to reshaping and enhancing the landscape, few matched the enthusiasm of the Norman invaders of 1066. With a mania for hunting, Britain's newest overlords depopulated large tracts of territory in the interests of blood sport. Dozens of hunting grounds, or 'forests', were designated, encompassing not just wooded areas but moorland, cultivated fields, and even whole villages, from which the occupants were banished under 'forest law'. Any animals which could jeopardise the chase were also dealt with with ruthless efficiency: sheep and goats, whose grazing could damage the forest vegetation, were removed, and unwanted dogs hobbled in a procedure known as 'lawing', which saw the claws from one foot lopped off with mallet and chisel. The Anglo-Saxon Chronicle for 1087 implies that William the Conqueror's focus was native game: 'Whoever slew a hart or a hind [male or female red deer] was to be blinded. He forbade the killing of boars even as the killing of harts. He loved the harts as dearly as though he had been their father. Hares, also, he decreed should go free.' Yet, William and his successors seemed happy to bring in, and protect, foreign quarry species.

This included the fallow deer. Like the indigenous red deer, fallow offered fabulous sport for the mounted hunter and hound by galloping away across the countryside. (The roe, Britain's other native deer, was far more skittish and a bit of a killjoy: its instinct was to hunker down in thick undergrowth at the least sign of danger, and it could even die of fright.) As discussed, small numbers of fallow deer may already have been present in Britain before the Normans; certainly, by the beginning of the twelfth century the species is known to have been well established. There's also a possible Sicilian connection here.

After a 30-year campaign, the Normans completed their capture of this Mediterranean island from the Arabs in 1091. Perhaps impressed by the parks of wild animals, including fallow deer, kept by Sicily's previous rulers, in 1129 King Henry I had 11 kilometres of wall built around his own estate at Woodstock, Oxfordshire, to which he introduced lions, leopards, camels and a porcupine. And fallow deer. According to the archaeologist Naomi Sykes, 'This collection, which is the direct ancestor of London Zoo, was not simply a frivolity; it was a metaphor for the Norman Empire, a statement that the Norman kings had power not only over the wild creatures in their possession but also over the countries from which the animals derived.' In addition to being far more manageable than red and roe – their scientific name *Dama* comes from the Persian for 'tame' – fallow thrived on poor quality land, so proved an immediate hit. By the 1300s, the deer had been stocked in some 3,000 parks across Britain; in England alone, these enclosures covered the equivalent of 2 per cent of the entire land area. The modern distribution of fallow deer, whose UK population probably exceeds 200,000 individuals, matches that of the medieval parks from which they escaped. (According to Charles Smith-Jones of the British Deer Society, fallow are remarkably loyal to their home areas and seem inclined to heft strongly to them.) Like other deer species – both native and introduced – the fallow is today regarded as a crop pest, an unwitting cause of vehicle collisions, and a potential carrier of disease from bovine tuberculosis to foot-and-mouth.

The common pheasant was already successful before its introduction to Britain, having colonised a swathe of Eurasia from the western Caspian region to Japan. As discussed, in Britain its bones first turn up at Roman sites, and historical documents – most of them written after the fact – indicate that pheasants were sometimes eaten as a luxury prior to the Norman invasion. For instance, in 1059, King Harold is said to have offered the bird as a privilege to the canons of Waltham Abbey in Essex, a gift deemed equivalent in value to a brace of partridges or a dozen blackbirds. In 1098, Radulfus, the Prior of Rochester, dispatched to his monks 16 pheasants (along with 1,000 lampreys, 300 hens, 30 geese, 1,000 eggs, 4 salmon and 6 bundles of wheat). A contemporary and perhaps more dependable record – a bursar's roll at Durham Priory dated to the reign of the Saxon king Edward the Confessor (1042–1066) – includes a purchase of one pheasant and 26 partridge. Pheasants may first have been kept in royal parks and forests, along with fallow deer, and their increasing prominence on banquet menus from the late twelfth century implies that they had by then naturalised. In 1251, Henry III ordered 290 of them for his Christmas feast, and by the late

1400s, pheasants warranted legal protection from the Crown. These early imports were in fact the 'Old English', or *colchicus*, subspecies from the Caucasus and lacked the distinctive white neck ring of the *torquatus* race, originating in China, which is these days released for shooting.

The pheasant is something of an outlier from this period in retaining a certain aristocratic association. The best explanation is that these poorly camouflaged, clumsy fliers have so far failed to get along in the British countryside, despite repeated reintroduction. Of an estimated 20 million poults (young birds) loosed annually, 90 per cent perish within the year. And not just from the shooting: most evade the guns only to be picked off by foxes or end up as roadkill.

Although pheasants might one day naturalise in Britain, there's precious little evidence for that so far. The same can't be said for what is without doubt the most impactful of all medieval introductions.

'It's quite a massive hill here, this site,' I said. 'And that's all just made for the rabbits, this big mound?'

'Well, no. The hill, I think, is natural. It's just that rectangular mound there that's made for them,' responded David patiently.

'Sorry. I was thinking the *whole hill* was a warren!'

'Oh. That would make it the world's biggest pillow mound, yeah.'

The aroma of wood smoke wafted in the chilly morning air. A distant chainsaw whined. We were standing at the foot of a steep grassy hillock, almost 100 metres high, upon which was broodingly perched a three-storey tower of limestone. Known these days as the Bruton 'dovecote' for its later repurposing by pigeon-fanciers, the original function of the pale-yellowish structure, which dates back to the 1500s, is a mystery. One theory has it as the prospect tower for the nearby Bruton Abbey – long since demolished on the orders of Henry VIII during his dissolution of the monasteries – offering the local aristocracy a grandstand view of the abbey's deer park. But neither doves nor deer, nor indeed the pair of Friesian cows munching contentedly near the base of the tower, had drawn me to South Somerset today.

No, I was interested in rabbits and in particular how and why these shy burrowing mammals from southwest Europe had been introduced to Britain, and then run amok. An important clue was offered by the pillow mound, a characteristic earthwork which to the trained eye shouts 'rabbit'. Clearly, my eye wasn't trained because Dr David Gould, a landscape archaeologist from the University of Exeter who had agreed to show me some, needed to point out the example that was right in front of us.

'You see that ridge coming down the hill?' he said. 'That's one.'

The British population of the European rabbit today numbers in the tens of millions and the species is now regarded as a worldwide menace. Yet the original bunnies were an ineffectual lot, hardly a patch on their vigorous descendants and quite unable to excavate their own burrows. This is where the pillow mounds came in. Created by piling up soil in long, low heaps, and encircled by ditches, possibly to deflect any floodwaters, these artificial structures provided a dry, soft and well-ventilated substrate into which the rabbits could dig. Some even incorporated stone-lined tunnels making life easier still for their feeble tenants. At the same time, pillow mounds concentrated the rabbits in one place for 'hunting'. If you could call it that. The phrase 'shooting fish in a barrel' comes to mind.

The pillow mound was a hallmark of the artificial rabbit warren, or 'coneygarth', from the Middle English *coning-erth*. *Coney, coning, conyng,* and sundry other derivations thereof, was the original word for the adult animal, the term 'rabbit' – from the French *rabette* – being reserved for juveniles. David had spent three years visiting 650 coneygarths across south-west England, from Cornwall to Wiltshire, racking up more than a thousand pillow mounds along the way. Little wonder he knew one when he saw it.

'Most are rectangular, like the ones here at Bruton,' he said. 'But you get circular ones, oval ones, cruciform ones. Just random, weird little ones.'

I suspected pillow mounds haunted his dreams.

Along with documenting the shapes, David was keen to understand just how conspicuous the pillow mounds were: 'If you were wealthy, you were expected to have access to these animals. It was kind of like the "in thing". But I wanted to know whether pillow mounds themselves, as visual components of the landscape, had a symbolic significance in their own right. Was it like parking your expensive car in the front drive to show off?'

In the event, David's field work revealed no clear pattern: pillow mounds were as likely to be tucked away behind a hill as to be sited ostentatiously on its slopes. It seemed that so long as the lord of the manor could offer distinguished guests fresh rabbit for dinner, whether or not the warren was visible from the manor house was of little concern.

As with other exotic imports, rabbits served multiple functions, offering meat, fur and status. Like pheasants and fallow deer, the association with elites can be traced as far back as the Romans who, elsewhere in their empire, prized rabbit foetuses, known as *laurices*, as a delicacy and reared the creatures (along with hares) in stone-walled pens called *leporaria*. The discovery of a fragment of rabbit tibia at Fishbourne, dated to the first century CE, suggests the species was brought to Britain during the Roman occupation, perhaps as a pet. But rabbits don't seem to have established: there's no Anglo-Saxon word for them and they don't get a nod in the Domesday Book. 'Coney culture' nevertheless persisted on the continent after the Romans left and, by the Norman period, rabbits had been added to the variety of smaller game that aristocrats would seek permission from the king to hunt under the right of 'free warren'. (Other free warren species – undoubtedly offering more sport – included fox, hare, wildcat, pheasant and partridge.)

Britain's current rabbit population dates to the second half of the twelfth century, with animals possibly brought by homeward-bound crusaders. At first, the rabbits were kept on islands off the south coast of England, the benign climate and lack of predators suiting these delicate mammals. Although there's some dispute about it, the earliest putative record dates to around 1135 when Drake's Island in Plymouth Sound was said to have been granted to Plympton Priory, *cum cuniculus* ('with rabbits'). In 1176, rabbits were being kept on the Scilly Isles, while on Lundy in the Bristol Channel, the tenant was permitted to take 50 a year between 1183 and 1219. One of the earliest allusions to mainland rabbit-keeping dates to 1235 when King Henry II presented ten live coneys as a gift from his park at Guildford.

Soon after the introduction of rabbits to mainland Britain, coneygarth escapees were turning up as pests on nearby arable fields, yet the species remained scarce during the early years. This rarity was reflected in the price, with a single animal costing the same as five chickens. Coneygarths were guarded and poachers subject to the full weight of the law. In England alone, 465 cases of rabbit theft are recorded between 1268 and 1551. Contrary to popular belief, peasants weren't always – or even mostly – responsible for rabbit-thievery. Break-ins were more often than not the handiwork of fellow landowners in a spirit of aristocratic one-upmanship. Warreners, who were tasked with ensuring the safety of their precious charges, had their work cut out. They constructed lodges and watch-towers to spot poachers, and fitted ingenious vermin traps to divert and capture stoats, weasels and other would-be predators. Trowlesworthy Warren on Dartmoor, which dates back to the seventeenth century, boasted 76 such traps. But not every rabbit predator was quite so unwelcome.

The ferret, a tame version of polecat which originated in North Africa and had been domesticated since the fourth century BCE, appeared in Britain from 1223, soon after the dawn of rabbit-keeping. Warreners co-opted the wiry carnivore to their cause, filing down its teeth and using it to flush bunnies from their burrows. Ferrets also formed an important component of the rabbit-poacher's toolkit, along with dogs and nets.

Against the odds, the rabbit population started rising, and by the fourteenth century supported a growing export trade in their furs; in 1305, for instance, 200 skins were shipped out of Hull, and by 1398 a certain Collard Chierpetit was granted the right to send 10,000 rabbit pelts to Holland. No fewer than 4,000 rabbits were served at the 1465 investiture of the Archbishop of York and, a century later, the Swiss naturalist Conrad Gesner noted that: 'There are few countries wherein coneys do not breed, but the most plenty of all is in England.' It wasn't until the late 1700s, however, that the wild population properly took off; rabbits had by then evolved into an altogether hardier proposition, able to capitalise on new rotational field systems, which provided a year-round supply of food.

The wholesale removal of weasels, pine martens, polecats, stoats, foxes and other predators by gamekeepers tasked with preserving pheasant and partridge, also indirectly benefited the rabbit whose ubiquity helped consign their high-class status to history. Rabbit farming continued in Britain right up to the twentieth century, with numerous large coastal semi-natural warrens in places like Cornwall and South Wales continuing to be protected from poachers, suggesting that the species retained a certain economic value until modern times. Nevertheless, this once-prized commodity fit for a king generally came to be dismissed as a pauper's ration, and at worst, vermin to be eradicated.

Rabbits have long had religious connotations, most likely anchored in the supposed proclamation of the sixth-century Pope Gregory that rabbits, or more precisely their foetuses, were fit for eating on fast-days. Plucked from the womb's watery environment, the reasoning went, laurices could be deemed honorary fish, not warm-blooded animals which would have been off limits. But this turns out to be a case of sloppy scholarship: the Pope never made any such decree. Instead, it was his contemporary and namesake, Bishop Gregory of Tours who had pronounced on rabbits, and merely to report on the practice of laurice consumption during Lent. Chinese whispers did the rest.

Writing in the 1990s, the archaeologists David and Margarita Stocker nevertheless detected an allegorical significance in 'defenceless' communities of rabbits being 'herded and managed like sheep' by a Christ-like warrener, before emerging 'from the ground to fulfil themselves'. Warming to their theme, the Stockers evidenced the deliberate, prominent and 'symbolically meaningful' placement of pillow mounds within monastic precincts at Sawtry Abbey in Cambridgeshire, Nun

Coton Priory in Lincolnshire and Croxton Abbey in Leicestershire. In his own documentary research on rabbits, however, David Gould has found little to bolster, or at least privilege, the sacred connection. 'In the medieval period it's basically the elites who first owned rabbits,' he said, 'and that meant both lay and clerical elites. In fact, when you look back through the records, warrens are more often linked to secular aristocracy.'

As it happened, Bruton had something to say on fish too (another reason David had suggested we meet here). We trudged to the crest of the dovecote-dominated knoll which gloried in the name of 'Lusty Hill'. Was this a reference to the renowned reproductive capacity of its former livestock? That was a question for another day. Passing two smaller pillow mounds on the summit, we descended the far side to a series of boggy depressions, the remains of ancient and overgrown ponds. 'I'm not an expert, but I think they're medieval and older than the pillow mounds,' said David. These artificial pools – fed by a stream which flows on to the River Brue – once supplied fresh fish to Bruton Abbey.

Whether or not rabbits were regarded as fish and kept and eaten for their ecclesiastical significance is unclear, but *actual* fish certainly were favoured by religious orders across continental Europe and, between the ninth and eleventh centuries, appeared ever more prominently on the table at Benedictine monasteries. The stricter regimes at Cistercian and Carthusian communities resisted even fish but later allowed the consumption of small quantities, or 'pittances'. The medieval period coincided with a massive expansion of marine fishing which targeted herring, cod and hake. While coastal communities enjoyed fresh catch, those living far inland had to make do with salted or dried fish. The elites, though, abhorred preserved fish and instead focused on locally caught freshwater species, like eels, which were trapped as they migrated up or downstream. The younger specimens were grown in stock ponds, either purpose-built or adapted from existing mill-ponds, moats and former river channels.

Meanwhile in Britain, the eating of fish also grew in popularity in the centuries following the Norman Conquest. The people of this island nation had always enjoyed access to seafish, but the continental pond culture centring on fresh-water varieties was nevertheless imported, as much for its prestige as for its nutritional benefits. Like deer parks and coneygarths, Britain's fishponds were first associated with the wealthy, and by 1300 could be found on the estates of clergy, aristocratic landowners and the Crown from Wiltshire to Yorkshire – the ones at Bruton being surviving examples. In time, husbandry techniques advanced enough to provide a steady food source and aquaculture became a widespread commercial enterprise, although freshwater fish retained its cachet. The sorts of fish best able to endure the warm, slow-moving, turbid and oxygen-poor water of medieval ponds were favoured. That meant roach, tench, chub, dace, perch, and especially pike and bream. To this roll-call of hardy natives would later be added a foreign fish whose origins are as murky as the waters it frequents.

The common carp is today among the world's most important food fishes. Three million metric tonnes are grown annually across 100 countries, equivalent to a tenth of all freshwater aquaculture production. It's easy to see why: the fish breeds and grows fast, tolerates a wide variety of environmental conditions and eats pretty much anything that can be sucked up by its telescopic mouthparts. The koi carp, a colourful variant developed in a mountainous region of Japan, is perhaps the world's most popular outdoor ornamental fish and almost as well travelled as its edible cousin. Meanwhile, the species supports an angling market which in Britain alone is worth close to a quarter of a billion pounds each year.

The ancestral common carp evolved close to the Caspian Sea around 2.5 million years ago and, taking advantage of the proliferation of waterways during warmer interglacial periods, expanded its range east into mainland Asia and west to the basins of the Black and Aral Seas. The European version of the common carp appeared in the Danube river some 10,000 years ago. And there the fish might have stayed were it not for its discovery by the Romans – keen aquarists – sometime in the first or second century CE. (Carp bones dated to that period have been identified at the site of a former Roman frontier fort near Iža in Slovakia.) Able to survive out of water and without food for prolonged periods, the carp were transported, possibly wrapped in wet moss or sacking, to the *piscinae* (reservoirs) of Italy as gourmet items and pets.

Carp however, live specimens at least, weren't present in Roman Britain, and don't feature in European pond culture until around the twelfth century. The first written reference on this side of the Channel comes from the kitchen accounts of King Edward III at Canterbury, dated to 1346, which show a carp and eight pike costing 22 shillings. The carp in this case was probably an imported specimen, because the fish doesn't seem to have been stocked in this country for another century. In 1496, *The Boke of St Albans* – attributed to Dame Juliana Berners, prioress of Sopwell nunnery – describes the carp as a 'deyntous [delicious] fisshe', and then in 1532 'Carpes to the King' appears in Henry VIII's Privy Purse expenses for that year.

Despite this, carp was historically less important in Britain than elsewhere, perhaps because by the sixteenth century improvements in navigation and ship technology were, for the first time, allowing exploitation of vast new shoals of marine fish from offshore Atlantic waters. The common carp nevertheless qualifies as among the first non-native fish to have naturalised in Britain and remains abundant in still and slow-flowing waters across England, with scattered populations in Wales and Scotland. The species is often accused of muddying the water as it ploughs river and lakebeds for invertebrates, fish eggs and other buried morsels. The resulting high water-turbidity stops light penetrating and interferes with photosynthesis, messing up food webs. But carp enthusiasts, of which there is a growing army in Britain, argue that recreational boating and other human activities are as much to blame.

A fascination for all things botanical, both native and exotic, also germinated within the monasteries and aristocratic households of Britain during medieval times; commercial horticulture can be traced to the thirteenth century, with enterprises in London and Oxford selling seeds in large numbers. *Husbandry*, a set of rules for estate management by Sir Walter of Henley published in 1280, states that imported corn-seed often outperforms home-grown counterparts, and this influential work may have encouraged the acquisition of foreign plants. By the late 1300s the Dominican friar and herbalist Henry Daniel was nurturing 252 sorts of herb in his garden in Stepney, London, of which 100 were non-native.

Plants were cultivated primarily for function not aesthetics, although the beauty of snapdragons, snowdrops and snake's head fritillaries – all of them apparently introduced during this period – is undeniable. Dill, coriander, summer savory,

black mustard, fennel, caraway and parsley were all condi-
ments whose use had declined after the Romans left but which
made a big comeback during medieval times. Hitherto
unknown species also arrived including saffron, a luxurious
yellow spice made from the dried stigmas of a crocus flower.
The plant originated in western Asia and is first recorded in
England in the fourteenth century. Used as culinary ingredient,
dye, perfume and aphrodisiac, saffron was famously grown
in East Anglia, its economic significance such that a major
centre of production, the Essex town of Walden, adopted it
as a prefix in the sixteenth century.

Horticultural introductions served other purposes. As its
name suggests, the leaves and roots of soapwort, a member
of the pink family native to the Middle East, contain natural
detergents. Appearing in Britain from medieval times, soapwort found use in the wool trade, washing not just woollen
products but the sheep from which they were derived; as
recently as the 1970s, extracts were employed to clean fragile
tapestries. Chasteberry, a type of vervain with purple flower
cones which originates in the eastern Mediterranean, was
used in monasteries to suppress libido among the acolytes,
and nuns stuffed their bedding with the aromatic leaves to
quash wicked urges. (The ancient Greeks prized the plant for
the reverse effect: women slept on it to enhance their fertility.)
The Aegean wallflower, meanwhile, was esteemed for the
fragrance of its vivid golden blooms, reminiscent of violets.
In its home range, the plant spreads over cliffs, and may have
first reached Britain stuck to building stone imported by the
Normans. It's still found clinging to ancient edifices from Bury
St Edmunds Abbey in Suffolk to Northumberland's Lindisfarne
Priory.

Almost every introduced plant offered some or other kind

of therapeutic function. Gout was treated with wall germander, a variety of mint; feverfew, in the daisy family, was a traditional painkiller; hollyhock, a laxative. Many plants were considered panaceas. Sweet cicely, a celery relative whose strong scent called to mind myrrh, was one of countless such 'cure-alls' and was used to remedy rheumatism, cleanse cuts and salve sore throats. It could relieve asthma, cure snakebite and promote sleep. Sweet cicely even stopped you farting. From time to time, serious mistakes could be made: to medieval midwives, the pretty yellow flowers of birthwort, a variety of clematis, resembled wombs – one shudders to imagine how they would know that – and they would administer its sap during labour to expel the placenta. It turns out that birthwort extracts are carcinogenic and may have killed thousands of women over centuries of misuse.

Such cases were rare however, and did little to disillusion medieval herbalists. Yet, in the fourteenth century there arrived in Britain a disease – caused by one non-native and apparently carried by others – which even sweet cicely would be powerless to prevent (although people gave it a go). It would help change the course of human history, disrupting existing power structures and kick-starting an era of empire building and world exploration to dwarf anything achieved by the Romans and Normans. And the unprecedented globalised trade and migration that resulted would turn a trickle of non-native species into a deluge.

4

New Worlds, New Invaders

'. . . They saw many kinds of trees and plants and fragrant flowers; they saw birds of many kinds, different from those of Spain, except partridges and nightingales, which sang, and geese, for of these there are a great many there. Four-footed beasts they did not see, except dogs that did not bark.'

The Journal of Christopher Columbus,
Tuesday 6 November 1492

In June 1348, a few days before the feast of St John the Baptist, a seaman came ashore at Melcombe Regis on Dorset's southern coast. Some say he'd been in Bordeaux, others suspected Calais, captured the previous August by King Edward III's forces. Whatever the truth, he was out of sorts.

More than likely he was running a fever and his joints ached. He may have been vomiting. In a day or so, boils in his neck, armpits and groin would swell and erupt in a mess of blood and pus. A week later, the sailor was almost certainly dead. The same fate soon befell others in his crew, and in time much of Melcombe's populace would follow them to the grave.

The town's merchants are said to have hushed up the calamity to protect trade in what ranks as one of history's most brazen cases of 'business as usual'. But by then the same tragedy was probably playing out at other harbours in southern England, as numerous mercantile and military vessels arrived from the continent, spilling out a cargo of people, goods and the plague. By 15 August, the contagion reached the port of Bristol, dispatching many of its citizens; some are said to have perished within hours of infection. A month or two later London was hit, losing up to half of its citizens. The spring of the following year saw the Midlands and Wales ravaged. By the time the disease abated in September 1350, few corners of mainland Britain had been spared. The poor were worst affected, but the disease killed off the wealthy too: early victims included three Archbishops of Canterbury and the Abbot of Westminster. Almost half of England's clergy and a quarter of its aristocracy would succumb.

The plague bacterium *Yersinia pestis* had been knocking around in Eurasia for millennia before strains evolved capable of annihilating humans on a massive scale. The fourteenth-century pandemic, which came to be known as the 'Black Death', was not the first of its kind – some 800 years earlier, an outbreak exterminated 30 million across the eastern Roman Empire – but it was among the worst. The variety of *Y. pestis* responsible emerged in central or eastern Asia during the early 1300s and followed overland trade routes west across the steppes to the Black Sea. In the autumn of 1346, while attacking the Genoese outpost of Kaffa (modern-day Feodosiya) on the Crimean peninsula, Mongols are said to have catapulted infected corpses into the besieged port. The diseased merchants took flight the following spring, carrying the sickness to Constantinople, Pisa, Genoa and Venice. With a foothold in these great trading centres, the plague propagated in all directions. When in 1353 it petered out in Russia, the 'Great Pestilence' had slain a third of all humans across the Middle East, Europe and North Africa. And it wasn't finished: the same plague strain flared up time and again over the next 400 years, famously returning with a vengeance to the seventeenth-century London of Samuel Pepys.

Thanks to records kept by diligent scholars, the route and chronology of the fourteenth-century plague's initial spread to Europe is well understood. Less certain is how and why it moved at such a pace. To this day *Y. pestis* is harboured in rats, marmots and many other ground-dwelling rodents, and transmitted between hosts by fleas. While the contagion can also be spread through the air, most modern cases of plague in humans occur when an infected rodent flea goes on to bite someone. One rodent in particular, the ship rat (or black rat), has traditionally been held liable for the devastating promulgation of the Black Death to Britain and the rest of Europe. And for good reason: it loves human company – even if the affection is unrequited – and, as its name suggests, is readily spread by marine vessels. Originating in the Indian subcontinent, ship rats are known to have been in Egypt some 3,000 years ago and reached British shores during the Roman occupation. For some reason the rodent disappears from the archaeological record during the Dark Ages, perhaps suffering from a decline in urbanisation and the cold, wet climate that characterised this period. By the time of the Black Death, however, ship rats and their fleas once again infested trade routes from Cairo to Cardiff. It's easy to see why the rats get the blame.

But some scholars are dubious, arguing that the rats themselves are as vulnerable as humans to lethal strains of *Y. pestis*, so make for inefficient carriers. In other words, they would have died from the plague faster than they could have disseminated it. This might explain why modern outbreaks, such as one in Glasgow in 1900, irrefutably caused by rodents and their fleas, seem altogether less catastrophic affairs than the Black Death: just 16 people died in the Glasgow event. An alternative theory now gaining credence is that the fourteenth-century plague was transmitted directly between people via their own human-specific lice and fleas. The idea is supported by recent mathematical simulations of medieval plague outbreaks in nine European cities for which good historical records are available. The research indicates that the pattern of plague transmission better matches spread by human parasites than spread by rats or air. Was the ship rat an innocent bystander all along? If so, we can remove at least one stain from the character of this much-reviled non-native.

What was the longer-term impact of Black Death? Again, there's much debate. At first, the fear of spreading the contagion curtailed the movement of people, slowing trade and collapsing Europe's thriving medieval economy. The massive loss of life, particularly among the poor and young, led to a labour shortage, boosting wages and empowering a hitherto subservient workforce; if the lord of the manor ill-treated his few remaining serfs, he could soon see them marching off to a more accommodating master. With a shortage of hands to harvest arable crops, forms of agriculture requiring little human input such as the farming of sheep and rabbits grew in importance, while the countryside emptied. There was a revolution in social mobility, with the lowliest of peasants feeling able to develop their talents and aspire to a higher station in life. People began questioning ancient belief systems as never before.

The plague was at first interpreted as divine punishment, prompting a surge in religious fervour as self-flagellation and other extreme acts of penance took hold. Yet waves of pestilence continued to sweep the land, taking innocent and guilty alike. Priests and other religious figures often bore the brunt as they tended to the sick, heard confessions and administered the last rites; their mortality rates were among the highest of any group. If devotion to God couldn't save them, what hope was there for everyone else? There's also the suggestion that the Church's reputation was further undermined by the lower moral and educational standards of new priests, monks and nuns hastily recruited to fill the vacancies. Faith in a fixed, pre-ordained world eroded, as power ebbed from established structures and new ways of thinking emerged. Human reasoning, values and experiences seemed more useful than religion in interpreting the present, past and future. Thus, the Black Death sowed the seeds of the Protestant Reformation, the Renaissance and a growth of scientific and rational thought which, by the fifteenth century, blossomed into a new age of technologically driven world exploration whose vision and scale would be unprecedented.

When it came to global maritime exploration, Columbus, Magellan, Vespucci and other famed adventurers from the fifteenth century were relative latecomers. Driven by fear, hunger or simple curiosity, humans have long sought to cross water, be it a woodland stream or a raging sea. And, every time, a hodgepodge of other organisms has joined for the ride.

The discovery, in the late 1990s, of 800,000-year-old stone artefacts characteristic of *Homo erectus* on Flores, in the Indonesian archipelago, tantalisingly suggests that we weren't the first hominids to build seagoing vessels for a spot of island-hopping, although it is possible *Homo erectus* may have just clung to rafts of vegetation and drifted there. We are, however, pretty confident that around 65,000 years ago representatives from our own species fashioned simple boats to cross the 150-kilometre-wide Torres Strait from New Guinea to Australia. From around 30,000 years ago, people began pushing further out into the Pacific Ocean, with the skilled seafarers of the Lapita culture migrating from Taiwan to colonise Micronesia, parts of Melanesia and western Polynesia about 3,500 years ago. Using better boats and navigational techniques, their descendants – often just called 'Polynesians' – later travelled to the remotest of Pacific islands, reaching New Zealand by about 150 CE (settling much later) and Hawaii in 400 CE.

Along with their language and technology, the adventurers brought plants for food and fibre. These included root vegetables, such as taro and yam; nuts and fruits like banana, breadfruit, coconut and *kukui* (candlenut); and *wauke*, the paper mulberry, whose bark made everything from bedsheets to fishing nets. Livestock came too, slung between the Polynesians' double-hulled canoes. Chickens and pigs provided meat, and dogs offered company. Many of these introductions naturalised in their new homes, and a few wrought profound ecological changes in the process. Uninvited passengers also came, including the Pacific rat, or kiore, which just like its

western cousin, the ship rat, was adept at smuggling itself
aboard maritime vessels. The kiore carries plague too,
although its greatest impact has been to deplete, and sometimes wipe out, indigenous bird, lizard and insect populations
each time it scampered onto hitherto rodent-free lands. Recent
genetic studies indicate that the kiore originated on Flores
(long after *Homo erectus* was there). The Indonesian island is
home to another rodent, twice the size of regular rats; unlike
its cosmopolitan cousins though the Flores giant rat is
restricted to the island, its dimensions perhaps thwarting any
attempt to slip unnoticed aboard a passing ship.

By the time the Black Death was casting its shadow over
Europe in the middle of the 1300s, the Polynesians had more
or less conquered the Pacific, perhaps even touching South
America. Arab and Chinese sailors had also got in on the act,
establishing complex maritime trading routes stretching from
the tip of Africa to Japan and doing their bit to spread new
species far and wide. But as impressive as these pioneering
explorations – and their impacts – were, they would be
dwarfed by the impending era of European expansion.

Some of the earliest transatlantic expeditions by Europeans
were driven by the search for new, more productive, fishing
grounds, with the quest for cod above all stirring the boldest
voyages. It was in pursuit of this nutritious and readily
preserved species that Norse explorers made their way to
Greenland in around 980 CE. Twenty years later they arrived
in Newfoundland, Canada, a territory referred to as 'Vinland'
in the Viking sagas. Five centuries on, the allure of cod, as
well as whales, was still drawing fishermen far across the
ocean, particularly sailors from the Basque region of northwest Spain and southern France. Along the way the eastern
Atlantic islands of the Canaries, the Azores and Madeira had

been discovered, settled and fought over by various European powers. As with the Polynesians in the Pacific, a host of new plants and animals, desired or otherwise, were introduced such as sheep, goats, cattle, asses, rats, grape vines, blackberries, sugar cane and rabbits. Many would have far-reaching consequences. For instance, the rabbit population exploded on the Madeiran island of Porto Santo in the mid-1440s after a female and her litter were released by its 'Captain', Bartolomeu Perestrelo. As Porto Santo's native flora was razed, much of its soil was eroded and washed away. Meanwhile weed species, which had also accompanied the Portuguese colonists, took over. Within a decade, the 15-kilometre-long islet was nigh on uninhabitable.

While the quest for bountiful new fishing grounds remained important, during the late fifteenth century a new urgency drove European navigators. For hundreds of years, Western civilisations had enjoyed – and come to expect – spices, porcelains, textiles, animal skins and precious metals from the Orient. In exchange, the Europeans exported timber, wool and citrus fruit. These and many other commodities went back and forth along a 7,400-kilometre-long network of land and sea routes known as the Silk Road which weaved between the great markets and port cities of Goa, Muscat, Zanzibar, Aleppo and Alexandria. The east–west commerce in goods, people and cultures had never been easy, with wars, natural disasters and piracy often interrupting trade and forcing re-routes. However, from the early fourteenth century, a period of singular instability beset the Silk Road arising from the disintegration of Genghis Khan's vast Mongol empire and culminating in the 1453 capture of Constantinople by the Ottoman Turks. Emboldened by advances in maritime science, unleashed in part by the Renaissance, Europeans began investigating new ways to the East which circumvented the increasingly problematic overland options.

In 1488, the Portuguese explorer Bartolomeu Dias became the first European to round the treacherous southern tip of Africa, and ten years later his countryman, Vasco da Gama, crowned the achievement by reaching India, thus connecting Europe to the Orient by sea for the first time. Meanwhile, as every schoolchild knows, Christopher Columbus was pondering a westward route to the Indies. Probably Genoan by birth (although this is debated), during the 1470s Columbus based himself in Lisbon, the centre of all the action, from where he and his brother Bartholomew peddled marine maps. Columbus learned open-ocean navigation and undertook numerous voyages under the Portuguese flag, visiting the British Isles, the west coast of Africa, along with the Canary, Madeira and Cape Verde islands. He later married none other than Filipa Moniz Perestrelo, daughter of the Madeiran rabbit-spreader Bartolomeu Perestrelo. Fast-forward a couple of decades and Columbus's interest in crossing the Atlantic had become an obsession that he was determined to make happen. But the recent success of Dias and da Gama had dampened enthusiasm for such an idiosyncratic project among likely sponsors in the Portuguese court, causing Columbus instead to seek and win backing in April 1492 from Queen Isabella and King Ferdinand of Spain.

Five months later, his trio of ships – the *Santa María, Santa Clara* (nicknamed *Niña*) and *Pinta* – set sail from San Sebastián de la Gomera in the Canary islands, the natural departure point for such a voyage. Recently acquired by Spain, the Canaries were located just off the coast of North Africa, enabling vessels to catch trade winds across the southern Atlantic. But the strategic position proved a mixed blessing since the archipelago would suffer centuries of biological incursions from all directions. The list of species whose harmful impacts are felt on the Canary islands to this day includes cats, rabbits, ferrets, rats, mice, goats, sheep, black-berries and elephant grass from the Old World, joined by invaders from the New, among them, prickly pear cactus, harpoon seaweed and California kingsnakes.

All this couldn't have been further from Columbus's mind as, early in the morning of 12 October 1492, after 37 days at sea, he caught sight of land, probably the Bahamian island of San Salvador. Looking back at him were indigenous Taíno people, from whom the Europeans went on to receive a warm welcome, along with parrots, cotton thread and other trinkets. Columbus headed home in January 1493 having, by all accounts, spent an agreeable three months cruising between other islands, including Hispaniola (modern Haiti and the Dominican Republic) and Cuba, and trading with 'Indians', as he called them, assuming he had reached the Orient. His souvenirs included pineapples, chili peppers, tobacco leaves and a turkey – all hitherto absent in Europe – not to mention six Taíno kidnapped as curiosities for the pleasure of Isabella and Ferdinand. Columbus completed three further jaunts across the Atlantic in his lifetime, taking with him pigs, wheat grains, grape vine cuttings, chickpeas, fruit stones and other Old World crops and livestock deemed essential for life on the new territory. The all too trusting Taíno were soon enslaved and forced on pain of death into a largely fruitless search for gold. Thus, from the earliest days, interest in alter-native routes to the Indies diminished in favour of a project to conquer and colonise.

Columbus is often said to have returned from his first transatlantic trip with maize, although this is debated; he would at least have seen the cereal's cultivation, for this was a staple of the Taíno people. Domesticated around 5,500 years ago in Mexico, maize – or 'corn' – was the Americas' answer to wheat and barley, and among the first New World crops to be widely introduced in Europe. By the end of the sixteenth century, maize, whose yields far outstripped its Fertile Crescent counterparts, was being grown across the Mediterranean region as far as Turkey and Egypt and had been planted in China. Today, about a trillion tonnes of the cereal are produced every year worldwide, almost as much as global rice and wheat production combined. Over time, as more of the Americas were reconnoitred, many other useful crops were discovered and brought home for cultivation in the Old World, including French beans, runner beans, tomatoes, cassava, chocolate, vanilla, pecans and pumpkins. The potato was a late addition to the list. Exploited in the Andes from around 7,000 years ago, the humble spud was first encountered by Spanish conquistadors in the 1530s but didn't catch on in Europe until new varieties were developed in the eighteenth century better able to cope with a temperate climate.

It's hard to overstate the positive economic and cultural impact of these and many other species brought back to Europe: just think of the significance of the potato to the Irish or the central role played by its close relative, the tomato, in Italian cuisine. And, generally, the introductions from the New World have proved benign. A mere handful of arrivals from across the pond loom large in the popular imagination as causing problems: grey squirrels, signal crayfish, ruddy duck and American mink. Few can disagree, meanwhile, that the New World got the short end of the stick in the great trading of organisms and technologies known as the Columbian Exchange, although 'short end of the stick' does little to describe the scale of the catastrophe visited upon the native people and habitats of the Americas and later colonised territories.

Perhaps the greatest blow was delivered early on in the shape of measles, mumps, smallpox, influenza and other infectious pathogens carried by Europeans, to which the indigenous people had no natural immunity. The statistics are sombre. Sixty years after Columbus's first voyage, the Taíno population of Hispaniola had crashed from three million to fewer than 500 individuals, the bulk of the deaths due to disease – although vast numbers were slaughtered by gold-crazed colonists. Similarly, within a century of the 1519 arrival in Mexico of Hernando Cortes, the number of indigenous Aztecs had fallen from a healthy 25 million to not much more than a million; at least five million perished from smallpox and other human-borne pestilences in the immediate wake of the Spanish conquest. A couple of decades later, a further 15 million Aztecs were killed by a typhoid-like haemorrhagic fever; recent research suggests this to have been a virulent strain of *Salmonella*, perhaps transmitted by domesticated livestock imported from Europe. No one can ever know the true figure, but it's been estimated that upwards of 56 million indigenous people across the Americas, Australia and New Zealand were wiped out by pathogens from the Old World. Syphilis, a bacterial infection, is one of the few diseases believed to have gone the other way; the earliest recorded European outbreak of 'the French disease' occurred in Naples in 1495 after infected troops invaded the kingdom on the orders of France's King Charles VIII.

The effects on newly discovered ecosystems would also be profound. Many were, and continue to be, caused by flora and fauna knowingly brought by Europeans. Negative impacts were felt soonest on the many remote islands upon which seafarers would release pigs, rabbits, goats, sheep and chickens as food for marooned comrades. A testament to this insurance policy is the tale of the eighteenth-century Scottish buccaneer Alexander Selkirk – the real-life inspiration for Robinson Crusoe. Marooned alone on a tiny island off the Chilean coast, Selkirk survived five years largely thanks to a thriving population of goats, descendants of animals left there 140 years earlier by Spanish colonists. But, as we saw on Porto Santo in Madeira, deliberate introductions such as these often flourished to the detriment of island ecosystems. Unintentional ones did as well. House mice colonised Gough Island in the South Atlantic a century ago, and today their giant-sized descendants regularly – and gruesomely – feast on the defenceless chicks of nesting albatrosses and petrels.

Although many introduced species established themselves in America, Australia and New Zealand right from the outset, the most notorious biological invasions date to the early nineteenth century, when political upheavals in the Old World triggered massive waves of human emigration. With steam-powered vessels cutting the cost and duration of intercontinental ocean voyages, immense numbers of people crossed the Atlantic from Europe to make a new life in the Americas. Millions also headed to Australia, New Zealand and Africa. And, repeating what should by now be a familiar pattern, these settlers weren't quite satisfied by what nature had to offer.

New World ecosystems were deemed lacking in certain 'superior' plants and animals from back home and needed to be 'improved'. That meant the planting of Old World crops, such as wheat and barley, the farming of Old World livestock, like cattle, sheep and goats, and the release of Old World fish and game for sport. Organisms were also brought over for pest control, as pets, or simply as mementoes of 'the Old Country'. In time, species introduction became both science and industry, as colonists – often the wealthier ones with time and cash on their hands – established 'acclimatisation societies' to import and try out a series of plants and animals in the newly acquired territories. Ecologists have been counting the cost ever since.

That only a small proportion of introductions succeeded is perhaps the saving grace of the acclimatisation project. Alas, the few that did naturalise have had an enormous impact. In the United States, one of the most widespread of all introductions, the European starling, was allegedly loosed for reasons which could be at best described as frivolous. In March 1890, the chairman of the American Acclimatisation Society, Eugene Schieffelin – a German immigrant and owner of a pharmaceutical business in the Bronx – decided to free 60 starlings into New York City's Central Park. The following year he released 40 more. It was all part of a scheme to populate America with every sort of bird mentioned in the plays of William Shakespeare. (If you're interested, the starling features in *Henry IV Part 1*.) The logic behind Schieffelin's plan is unclear: some say he hoped the civilised chirps of Old World birds could tame the city's uncouth masses. Whatever the truth, the starling is a fast-breeding and adaptable species, and that founding population of 100 birds multiplied and spread.

By the 1920s, starlings were in Florida and 20 years later had reached California, some 4,000 kilometres from their site of release. Today, an estimated 200 million of them are found across the continent, making them North America's most abundant bird. The species is blamed for eating $750-million worth of agricultural produce each year in the United States, often targeting enriched grain in livestock feedlots, as well as orchards. The sheer scale of starling gatherings – a single murmuration sometimes comprises tens of thousands of birds – poses a health hazard, with droppings contaminating water sources and passing on transmissible gastroenteritis virus to farm animals. The flocks have been known to bring down aircraft: in 1960 an Eastern Air Lines passenger plane crashed into Boston harbour with the loss of 62 lives after starlings gummed up three of its four propeller engines.

The United States Department of Agriculture has long conducted a war against starlings and other avian pests, including fellow non-natives such as blackbirds and crows, shooting, trapping and poisoning millions annually. Private citizens are meanwhile encouraged to apply detergents to starling roosts, spoiling the birds' insulation and causing them to perish in the cold. In recent years, the authorities have even used drones to scare them off. Starling advocates, on the other hand, bemoan the carnage and point to the free pest control services the insect-guzzling birds offer. If it were possible, there was, however, a still more disastrous Old World introduction during the nineteenth century, one that, unlike Shakespeare's birds, would lack any Stateside support.

Etienne Léopold Trouvelot: this French artist, astronomer and amateur naturalist may have wished to be remembered as the enlightened polymath he surely was. Instead, he has gone down in history as the author of a biological blunder of colossal proportions. In the 1860s, Trouvelot took it upon himself to bring back to his home in Medford, Massachusetts, the eggs of the gypsy moth. He was interested to discover whether the insect could be farmed for its silk. It could not. By the time Trouvelot got his answer, some very hungry caterpillars had wriggled free, their progeny proceeding to ravage forests up and down the eastern seaboard of the United

States. A century and a half on and the annual costs associated with the gypsy moth – which attacks 500 kinds of tree and shrub – exceed $3 billion across North America, putting it among the most damaging invasive insects on the continent. As with starlings, an arsenal of control techniques has been deployed against the gypsy moth, mostly in vain. A small mercy is that the moth targets deciduous trees, leaving evergreens alone. Also, the adult female is flightless, which reduces the pace of natural spread. Unfortunately, a new strain of the gypsy moth whose females *can* fly and which *does* attack evergreen species, has now arrived on the west coast of America, having apparently hitched a ride on ships from Asian ports.

Southern hemisphere colonies have also received many misguided introductions, the archetypal example being the release of European rabbits in Australia. They were initially brought as food with the First Fleet in 1778. Then, during the 1820s, small numbers established themselves on Tasmania after wild animals were introduced for hunting. But rabbits really got going in 1859. On Christmas Day of that year a farmer, Thomas Austin, released a dozen or more into Barwon Park, 115 kilometres east of Melbourne, Victoria. These had been brought from Liverpool on the brig *Lightning* (these details matter to the Aussies). A mere eight years later, the visiting Duke of Edinburgh was able to shoot 416 rabbits in the space of a few hours, and within a century close to a billion had colonised most of the continent, despite the best efforts of bounty-hunters and several thousand kilometres of rabbit-proof fencing. Dubbed 'chainsaws of the Outback', the Australian rabbits were a far cry from the frail creatures first brought to Britain in the twelfth century, ravaging crops, out-eating sheep and stripping tree seedlings. They were blamed for deforestation and desertification on a massive scale, and for supplanting native marsupials, such as the southern hairy-nosed wombat, the brush-tailed bettong and the greater bilby, and precipitating the extinction of the lesser bilby. It was the same in New Zealand, which saw rabbit numbers explode from the 1860s after a succession of releases. In 1936, settlers in Tierra del Fuego, Chile – who had clearly not been paying attention – also liberated rabbits; within two decades, the four original animals spawned a population approaching 30 million. They remain a serious pest of farmland and forestry to this day.

Many undesirable plants also travelled from the Old World to the New. When Europeans brought their millennia-old agricultural system to the colonies, the usual rag-bag army of weeds tagged along for the ride. Some did so quite literally: horehound, chamomile, sida retusa and capeweed are just a few of the unwanted plants which reached Australia as seeds clinging to the wool of imported sheep. What's more, the warm, dry climate in many colonised regions – such as central Chile, California, South Africa and southern Australia – matched the Mediterranean where an abundance of these plants had originated. With perfect growing conditions and a lack of natural enemies, the introductions often dislodged indigenous species and proved far more invasive than back in Europe. 'The endemic productions of New Zealand, for instance, are perfect one compared with another,' wrote Darwin in 1859 in *On the Origin of Species*, 'but they are now rapidly yielding before the advancing legions of plants and animals introduced from Europe.' Likewise, in 1881 the botanical artist, Marianne North, travelling to Tasmania in search of the unfamiliar, was disappointed, recording in her journal, that: 'It was far too English, with hedges of sweet-briar, hawthorn, and blackberry, nettles, docks, thistles, dandelions: all the native flowers (if there were any) were burnt up.'

Plenty of these and other problem plants were deliberately introduced. Gorse was among the more notorious of these, particularly in New Zealand where Darwin noticed it during a visit in 1835. Although its seeds may also have arrived as a contaminant of agricultural cargo, the species was intentionally brought from Britain by nineteenth-century settlers for hedging. With defensive spines deterring would-be grazers, gorse spread out of control, invading upland pastures and shrubland, and altering native habitats. Setting fire to the stuff just made things worse: numerous fresh seedlings rose from the ashes. By the 1980s, gorse had covered 3.5 per cent of New Zealand's land area. The plant, which thrives in poor soils, was also cultivated in the Americas, southern Africa and the Indian subcontinent, and is today considered one of the world's worst weeds.

Even when not laden with crops or livestock, ships helped spread non-native species. Until modern water ballast systems were developed, merchant vessels would be weighed down with soil, logs, rocks, gravel, rubble, chunks of iron and all manner of heavy objects to reduce the risk of capsizing. When cargo was ready to be taken aboard at a port, this solid ballast was tipped in great heaps on the quayside. (Local rules prohibited disposal in the harbour itself due to the costs of dredging.) Any plants, invertebrates and other soil-borne organisms in the dumped material then had a fighting chance of colonising the new location. Earthworms were notable beneficiaries of this invasion vector; so were beetles, which comprised around 90 per cent of all insects introduced to North America before 1820. Similarly, it's thought that the European fire ant, which has now all but displaced native counterparts in the Tifft Nature Preserve in upstate New York, arrived in ship ballasts at the area's former coal docks. Many hitch-hikers have been shown to have originated in southwest England and Ireland, reflecting the fact that ports in this region represented the last chance to scoop up material before the Atlantic crossing.

These ballast aliens, as they are sometimes known, were however generally short-lived in their new surroundings, and few strayed far. In Philadelphia, for instance, almost none of the 81 types of plant found close to the harbour during the late nineteenth century now occur in the city. Purple loose-strife, an attractive Old World perennial noted for its vivid mauve flower spikes and preference for damp habitats, proved a rare exception. The plant reached the eastern seaboard of North America sometime around 1800 in soil ballast (although it may also have travelled in wool imports and was later intentionally brought for ornamental and medicinal purposes). After gradually spreading along canals and irrigation systems for more than a century, purple loosestrife began to proliferate at breakneck speed in the 1930s and has since smothered thousands of hectares of temperate and boreal wetlands across the United States and Canada. Marine species picked up from European shorelines could also be spread in solid ballast; serrated wrack seaweed and the common periwinkle, now regarded as invasive on the east coast of North America, turned up in Nova Scotia, Canada, at least 150 years ago, after vessels from Scotland and Ireland dumped their ballast in exchange for timber.

In return Britain received its own assortment of soil-borne introductions, much to the excitement of Victorian and Edwardian naturalists who would take to scrambling over harbour-side ballast heaps in search of the exotic. In 1886, botanists recorded 67 varieties of foreign plant near Cardiff and Penarth docks, including Britain's first ever wild record of Japanese knotweed; and in Scotland's coal-exporting towns of Burntisland and Charlestown some 150 non-natives were found between 1820 and 1919. Ballast aliens were also reported from docks in Plymouth, Bristol, London, Birkenhead and Newcastle. Today, though, barely a dozen of these species persist, and most of those have also been intentionally introduced from time to time. One example is the American smooth cord-grass, first reported in Southampton Water in 1816. There's also the lesser swine-cress, a South American member of the cabbage family, and pineappleweed, a type of daisy from Asia which, within 30 years of its 1871 appearance, had spread across London. A fair number of plants from the European continent have also hitched to Britain in ship ballast including figs, henbane, branched horsetail and a variety of thistle found only in Plymouth. The phenomenon began to wane at the turn of the twentieth century with the adoption by ships of water ballast technology, although this would turn out to be a far more significant vector of invasive species.

In the early centuries of European exploration, ten times as many species of plant, animal and pathogen moved from Europe to North America than vice-versa. Australia, New Zealand, southern Africa and South America received more than their fair share of introductions. The marked disparity in flow of biological invaders – from Old World to New, northern hemisphere to southern – may go some way to explaining how a handful of human colonists managed to conquer the world. That's according to the late American professor Alfred W Crosby, who in his influential 1986 book, *Ecological Imperialism*, writes: 'Perhaps European humans have triumphed because of their superiority in arms, organization, and fanaticism, but what in heaven's name is the reason that the sun never sets on the empire of the dandelion?' Crosby argues that the colonials unwittingly benefited from an army of associated organisms, or 'biotic allies', which swept aside indigenous fauna, flora and even people, creating 'neo-Europes' in the process. But what made Old World species so dominant in the first place?

One theory is that their longer association with people better equipped them to anthropogenic disturbances; modern humans settled in Europe at least 100,000 years ago, but only reached Australia around 65,000 years ago, and the Americas 20,000 years ago. Hawaii and New Zealand, both badly affected by harmful introductions, were human-free until as recently as 400 CE. Moreover, the succession of Ice Ages across northern Europe forced many organisms to coexist more closely with people in climatic refugia, those pockets of warmer, habitable territory to which all and sundry retreated as temperatures dropped. Finally, in the last few millennia, a suite of plant and animal pests had evolved to capitalise on the sort of pastoral and arable agriculture developed in Europe. Thus, it seems that when Europeans ploughed up American soil, or built their first Australian settlements, European species tended to jump in first.

However, it wouldn't all be one-way traffic. As the age of global exploration dawned, the Old World itself would receive some new troublemakers. And, just like many of those running amok across the empire, some of the worst would be brought by design.

5

The Empire Strikes Back

'The chief aim of acclimatisation is not to substitute a foreign animal for an equally satisfactory native one; but, where and if it is possible, to fill up gaps in the home supply by good things from elsewhere. Another object is to replace an inferior native species by something incontestably better from abroad.'

'New creatures for old countries',
Quarterly Review, Charles J Cornish, 1900

It was just a bouncing flash of red-brown. An arched back. A white wisp of a tail. Dog-sized, but unmistakably a deer. Seconds later the animal had gone, plunging into a thicket of brambles bordering the village green at Santon Downham. I experienced that glow of satisfaction which accompanies any minor achievement: another species ticked off. For one of the country's more plentiful non-native mammals, the Reeves' muntjac had been darned tricky to spot. I drove straight back to the Forestry Commission office I'd just left to tell James (a false name due to the sensitive nature of his job). This seemed only fair since, as one of Thetford Forest's five wildlife rangers, he had given up much of his morning driving me around in a fruitless quest for Britain's smallest wild deer. James greeted the news with a smile.

I had clambered into his Toyota 4x4 at six that gloomy June morning. The cab smelt of bug-spray and a dog, whose lurking presence in a rear compartment of the vehicle was betrayed by the occasional whine. 'He's a Labrador cross German Wire-haired Pointer,' said James. 'Looked like a black Lab pup when he was born, but then he started to get whiskery.' This canine was both companion and essential part of the deer stalking toolkit, helping to sniff out unseen animals, and then tracking them down after the kill. 'Even with a good chest-shot in the heart and lungs, they can often lurch forward and run with the adrenaline,' explained James, 'and they may go 50, 60, 80, 100 yards, possibly into thick undergrowth. The dog is trained to follow the blood trail, and they will find you that deer.'

The ensuing two hours had been spent steadily touring many a kilometre of conifer plantation, interspersed here and there with patches of open heathland or copses of broadleaf. Several times a muntjac, not far off, tantalised us with its high-pitched bark (reminiscent of a crow's caw). But James had warned that late spring, with the bracken already waist-high, was a difficult time of the year to detect an animal which, full-grown, reached little more than 50 centimetres at the shoulder. And so it proved.

Beyond the odd pheasant, hare, squirrel and, somewhat incongruously, a domestic cat – all of which prompted false alarms – the forest appeared devoid of animal life. It was hard to believe that James and his colleagues, tasked with controlling the deer population in the interests of 'crop and biodiversity protection', every year culled well over a thousand muntjac. Thetford Forest's 20,000 hectares, which straddled the Norfolk–Suffolk border, were also home to good numbers of roe, and a smaller population of red and fallow, each species using the forest in its own way. Red and fallow were primarily bulk grazers, happy to munch on grass, if that's all there was, but also partial to saplings and herbs, as well as to crops in the surrounding arable land. The smaller muntjac and roe, by contrast, were selective browsers, homing in on more nutri-ent-rich items, such as young woodland coppice and wildflowers.

In high enough numbers, any sort of deer could cause serious economic damage, as well as harming biodiversity. A recent study of woodlands with dense deer populations found a 68 per cent reduction in understorey foliage, resulting in a loss of breeding habitat for ground-nesting birds like nightin-gales, along with butterflies and many other insects. Deer are involved in up to 74,000 road traffic collisions across the UK every year, so they hurt us too. Around 700 people are injured, and up to 20 killed, in these crashes annually. The roads around Thetford Forest, including the A134 which passes through its centre, are among the country's worst blackspots for such accidents.

James, therefore, had plenty of reasons to want to control deer on his patch, yet his love of them was obvious. 'A big part of my job is culling, but I do have a great respect for them and interest in them, which some people might find ironic,' he said. This affection even extended to non-natives like the muntjac.

'If you could turn the clock back, you would, but we are where
we are. Muntjac have been around for decades now, so it's more
a case of managing them with due respect to them as a deer
in their own right.' He paused before adding, 'It's not their fault
at the end of the day that they're here.'

In fact, all the deer in Thetford Forest were relative
newcomers. Although native to Britain, both roe and red had
long ago been exterminated in this part of the country. And
it was people who then brought them back. During the 1880s,
several pairs of roe were imported from Württemberg in
Germany and kept at a nearby estate from which they later
escaped. The presence of red deer, meanwhile, is a legacy of
the rarefied pastime of 'carting': from the nineteenth century,
stags and hinds sourced from Scotland or overseas would be
taken by wagon into the nearby countryside, released and
then tracked down using dogs, often over great distances.
Generally, the deer would survive the outing. At the end of
a day's sport the hounds would be kept at a safe distance as
the unharmed, but presumably terrified, quarry was roped,
loaded back into the cart and taken home. By the time carting
was dying out in the 1950s, small numbers of red deer were
living wild, having eluded the hunters. Thetford's fallow deer
population seems to have originated from nearby deer parks
at Ampton, Ickworth and particularly Livermere whose herd
was disbanded during World War I. (There were, in fact, small
and localised numbers of another non-native deer in the
vicinity of Thetford Forest: the sika. This Asian import doesn't
yet cause as many problems here as elsewhere in Britain –
notably on acid soils in Dorset, northern England and western
Scotland – where large herds trample farm crops, and damage
tree trunks by gouging them with their antlers, and, as
mentioned, interbreed with red deer.)

James was keen to stress that nothing was wasted, and back at the Forestry Commission compound, he showed me 'the larder'. This small, refrigerated outbuilding contained several dozen deer carcasses, the fruit of a week or so's culling activity in the forest. Hanging in neat rows, the animals had been prepared for sale to game-dealers according to strict criteria, with head, limbs and internal organs removed, ribcage sawn open, coat left on. By all accounts, muntjac venison is among the best. That's just as well because there won't be a shortage any time soon.

Despite intensive efforts to control the species across the country, the muntjac population currently stands at anywhere between 128,000 and two million, depending on whose study you believe. The wide variation in estimates is due to different assumptions for the average muntjac density in a typical patch of woodland; in truth, this small, secretive animal is virtually impossible to count. Nevertheless, in the words of one of James's colleagues, 'They're everywhere. Kick a bush and a muntjac will come out.' Judging from my trip to Thetford, it seems I'd been kicking the wrong bushes.

But how did this small, primitive deer, native to the wooded hills of southeast China, end up raiding suburban gardens in the British home counties? The story begins more than a century ago in one of the smarter districts of London.

At six o'clock in the evening of Thursday 23 March 1893, George William Francis Sackville Russell, the 10th Duke of Bedford, and one of the richest men in Britain, was discovered dead on the dining room floor at his residence in Belgravia. He was 40. Famed for his appetite and suffering from diabetes, his premature demise was not unexpected. Sackville had enjoyed ten years as Liberal Member of Parliament for

Bedfordshire, and his passing was keenly felt in the town of Tavistock, Devon – much of which he owned – whose agricultural tenants had recently benefited from several generous reductions in rent. The late duke was, though, something of a recluse, and said to have dismissed any female servant encountered after noon. Perhaps he had cause to be gloomy: after all, his father had shot himself just two years before.

Sackville left a wife, but no offspring, so his title and vast estate-holding passed to his younger brother, Herbrand Arthur Russell. Like his sibling, the 11th Duke seems not to have been a people person. Born in 1858 and home-schooled, he reputedly suffered from pathological shyness, manifesting itself in an autocratic exterior. Despite these qualities – or possibly because of them – the young Herbrand excelled as a professional soldier, seeing action with the Grenadier Guards in the 1882 Egyptian campaign, and going on to serve in Calcutta as *aide-de-camp* to the British viceroy Lord Dufferin. While in India, Herbrand met and married Mary Tribe, the daughter of an archdeacon. In 1888, perhaps anticipating a quieter life, the couple returned to Britain, leasing Cairnsmore, a shooting estate in the Scottish Lowlands of Dumfries and Galloway. Soon after his arrival, Herbrand strip-burnt most of the land in the interests of grouse-hunting and began casting about for exotic animal species to populate it. In so doing, he would lay the groundwork for what would be described as 'the largest of all experiments in animal acclimatisation in this country', a project which would absorb him for the rest of his life.

As we have seen, during the nineteenth century acclimatisation societies were established by British and other European settlers in far-flung colonies seeking to improve their surroundings with flora and fauna from back home. In fact, the origins of acclimatisation lay in the bringing of useful new species *to* Europe, and this had got going in a formal sense with the 1854 founding in Paris of La Société Zoologique d'Acclimatation. Its chairman, the zoologist Isidore Geoffroy Saint-Hilaire, wanted to 'people our fields, our forests, and our rivers with new guests; to increase and vary our food resources, and to create other economical or additional products'. Geoffroy Saint-Hilaire was influenced by the evolutionary theories of his countryman, Jean-Baptiste Lamarck, and believed that animals could be forced to adapt – or acclimate – to a variety of different environments. 'We have given the sheep to Australia; why have we not taken in exchange the kangaroo – a most edible and productive creature?' asked Geoffroy Saint-Hilaire, as he and his son, Albert, put the theory to test in a purpose-built zoo stocked with elephants, kangaroos, hippos and camels. The 'science' of acclimatisation often boiled down to simple trial-and-error, the zoological equivalent of throwing things at a wall and seeing what stuck – or 'salted', in the jargon of the time.

Regional societies soon sprang up around France, Germany and across the English Channel, the year 1860 seeing the inauguration of the UK's own version. The British initiative was led with vim by the doctor, naturalist and writer, Frank Buckland. Raised in a household where deep-fried mice, squirrel pie and ostrich weren't strangers to the menu, Buckland was noted for an open mind when it came to food, and indeed was inspired to found his society over a now-legendary dinner of eland, a type of African antelope, at the Aldersgate Tavern in London. As on the continent, Buckland and his fellow acclimatisers sought to procure for Britain a range of exotic wildlife for culinary and ornamental reasons. Belying the wide remit suggested by its title, the Society for the Acclimatisation of

Animals, Birds, Fishes, Insects and Vegetables within the United Kingdom focused on herbivorous animals, reasoning that they would be easier to establish than predators. The species selected were also ones that offered good sport, be they catfish from Austria, quail and bison from North America, wombats and Murray cod from Australia or barasingha deer from India. The project proved short-lived, however, reinventing itself as the Acclimatisation and Ornithological Society in 1866 and crumbling altogether four years later as members bickered over ownership of salmon eggs and Chinese birds.

If there's one message to be had, it's that the procurement, display and release of non-native species is a pursuit long favoured by Britain's elites. Even before Buckland's movement got going, the nineteenth century had already seen Edward Stanley, the 13th Earl of Derby, pioneer the captive breeding of Hawaiian geese, passenger pigeons and budgerigars at Knowsley Hall near Liverpool. Likewise, during the 1840s the naturalist-explorer Charles Waterton had loosed specimens of the little owl, collected in Italy, at his Walton Hall estate in Yorkshire. (Waterton's birds were never seen again, although within a few decades the species was flourishing in Britain thanks to subsequent liberations by his fellow aristocrats.) But, in acclimatisation, a craze coinciding with the publication of Darwin's theory of evolution, a gentleman's hobby was elevated to a science. And, although Buckland's society was dismissed as amateurish – typified by doomed experiments to cross Indonesian sambar with British red deer – its underlying philosophy that nature was there for humans to improve upon persisted.

Late nineteenth-century Britain thus saw the emergence of 'paradises', as preserves dedicated to acclimatisation were then known. These attracted the attention of the *Quarterly*

Review and other leading periodicals of the time, and included Haggerston Castle, near Berwick-upon-Tweed, which, despite what was described as a 'somewhat inclement' climate, had a thriving population of wallabies and American bison, along with wildebeest, emus and ostriches, and Japanese apes. Haggerston's owner, the entrepreneur and botanist Christopher Leyland, is now best known for nurturing the infamous cypress, a vigorous hybrid of Nootka and Monterey conifers, which bears his name. Then there was Sir Edmund Loder's 115-hectare Leonardslee estate in Sussex, whose very 'un-English landscape' boasted 'innumerable foreign species', among them Patagonian cavies, whose flesh was likened to 'coarse rabbit', and the 'largest numbers of American turkeys in one place in England'. Leonardslee was noted too for ante-lopes, gazelle and red kangaroo. A colony of 'industrious' North American beavers, meanwhile, displayed 'ingenuity' and 'harmonious co-operation', as 'mobs' of wallaby lay about on their sides basking in the sun.

Similar experiments were occurring across the Irish Sea at Lord Powerscourt's estate in County Wicklow, where sika deer, introduced from Japan in 1850, 'flourished in the most wonderful manner'. By contrast, the sambar that Powerscourt acquired a few years later 'all died of chills and dampness'. (Wallabies have generally fared a little better in Britain, with small colonies persisting to this day on the Isle of Man as well as on Inchconnachan island in Scotland's Loch Lomond.) But no paradise would approach the scale and scope of the extraordinary collection which Herbrand Russell would begin contemplating on his ascendancy to the dukedom in 1893.

The 11th Duke of Bedford was serious about his public duties. Despite turning down both cabinet and colonial appointments, he was active in the House of Lords, a proud lord lieutenant of Middlesex, Holborn's first mayor, as well as chair of Bedfordshire County Council for 33 years. During World War I, he acted as colonel-commandant of a training camp at Bedfordshire's Ampthill Park. Yet, estate management was his primary concern, along with the opportunity to indulge his passion for natural history. He may have inherited some of this from his late father, Francis Russell, the 9th Duke, who had himself dabbled in acclimatisation, but Herbrand's fascination was certainly stimulated by positive boyhood encounters with Frank Buckland, and Thomas Henry Huxley, famed as Charles Darwin's 'bulldog'. Among the finest of Herbrand's glittering portfolio of inherited property, encompassing tracts of Cambridgeshire, Devon and central London, would be his new home: Woburn Abbey. The estate, set amid gently undulating Bedfordshire countryside, had belonged to the Russells for almost 350 years following the dissolution of a twelfth-century Cistercian monastery on the site. While the duke would surely have appreciated the historic house, remodelled in Palladian style by successive leading architects and chock-full of art treasures, not to mention the gardens magnificently landscaped by the likes of Humphry Repton, his real interest lay in Woburn's 1,200-hectare deer park.

The tradition of deer management at Woburn stretched back centuries, so it was unsurprising that these stately mammals – long associated with the aristocracy – would be a core focus for its new, zoologically minded incumbent. But while the herd had, until now, been restricted to native red and the honorary native fallow, Herbrand had his sights set firmly on the more exotic. In his first ten years at Woburn, he sourced over a thousand 'foreign' animals from private collectors, or through exchanges with continental zoos (someone of his position didn't stoop to collecting the animals himself). He bred many more, with his wife, Mary, dutifully maintaining a register of all births, deaths and purchases.

New and delicate arrivals were kept under close observation in sheds and yards. The hardiest among them would graduate to more spacious paddocks, before a few were permitted the run of the whole park and neighbouring woods. The method seemed to work as, by the turn of the twentieth century, a 300-strong herd was living in 'semi-wild condition'. Of the 40 or so species then being trialled – representing nearly all known kinds in the world – nine were said, by the duchess, to 'survive an ordinary English winter in an open park with no more feeding than is usually given to red or fallow deer'. Early successes included axis from the Indian subcontinent, Japanese sika and the Caucasian red. American elk and reindeer, by contrast, were described as 'wholly incapable of acclimatisation'. By 1903, the overall deer population had stabilised at around 700 animals, parasitic diseases preventing the herd growing any larger. Tellingly, the 1903 figure doesn't include the Reeves' muntjac which had, by then, skulked off the books. 'They creep about in the thickets, and are hardly ever seen except when the coverts are beaten,' wrote one visitor at the time. 'There are a number at Woburn, but no record can be exactly kept of the rising numbers. They seem to be hardy and healthy.' Already, it seemed, Herbrand's most successful 'experiment' was abroad in the English countryside.

Even the origins of its common name are confusing. The Reeves' bit, that is. ('Muntjac' indisputably derives from *mencek*, a word used by the people of western Java for 'small deer'.) The mammal is invariably associated with John Reeves, an inspector of tea with the British East India Company, who was based at Canton in China from 1812. When tea-inspecting duties didn't get in his way, Reeves, like others of his ilk, devoted any spare energies to the collection and documentation of zoological and botanical specimens from the region, many of them new to Western science. He was, for instance, noted for the multitude of ornamental plants shipped home to satisfy the appetite among European gardeners for all things oriental, and also specialised in Chinese fish. However, one species John Reeves *didn't* discover was the Reeves' muntjac. That honour instead fell to his son. Not only was he also called John, but, like his father, he was a tea inspector in Canton and inveterate naturalist. It's easy to see where the confusion arises.

The career of the Reeves' muntjac in Britain didn't get off to a promising start: a pair first presented to London Zoo in 1838 died within three months. Interest in this unusual little animal nevertheless supported a steady nineteenth-century trade in specimens among zoos, deer parks and private animal dealers in Britain and abroad. Likewise, the related but much larger Indian muntjac, originating in parts of Indonesia, Malaysia and Thailand (but *not* India), was also popular. (For some reason, the Indian variety is often spelled 'muntjak'.)

Herbrand Russell came late to the party, starting off in the 1890s with several specimens of Indian muntjac, which he procured from London Zoo and released at Cairnsmore. These animals were later transferred to Woburn, where their numbers were supplemented with Reeves' muntjac. The duke struggled to get either type going in captivity and loosed them all into nearby woods in 1901. Soon after, however, both deer started breeding in the wild, within and without the walls of the grand estate. All seemed fine until, according to one legend, an Indian muntjac buck killed the terrier of the

Woburn resident rat-catcher (what muntjacs lack in stature is offset in feistiness and a pair of razor-sharp upper canines). By way of retaliation, Herbrand allegedly ordered the destruction of all Indian muntjac within the park. This story appears to have been false, however, because during the 1930s, the duke would donate 16 specimens of this deer to a new zoo he helped found at Whipsnade, 19 kilometres south of Woburn. Nevertheless, Indian muntjac eventually petered out in this country. Its Chinese counterpart, by contrast, was only just getting started.

The Reeves' muntjac had managed to survive in small numbers within, and close to Woburn, as well as in the grounds of Walter Rothschild's Tring Park in Hertfordshire, and were also donated to Whipsnade. Then, from the 1940s, the species started popping up in Oxfordshire, Warwickshire, Northamptonshire, Kent and East Anglia. (For years, the British muntjac were believed hybrids of the Indian and Reeves' but this has been disproved.) Like other deer, the average dispersal rate of muntjac is about one kilometre per year, so its appearance at more distant locations probably resulted from deliberate releases. Unlike more nervous counterparts, such as the roe, which can literally die of fright when handled (a condition known as 'post-capture myopathy'), the muntjac travelled well: they could be trapped, put in a box and driven halfway across the country, only to go bouncing happily off into the undergrowth. Such translocations, eventually outlawed during the 1980s, continued through much of the twentieth century, including at Norfolk's Elveden Estate, which may explain the presence of muntjac at nearby Thetford Forest. Today, the species is recorded across most of England, half of Wales and sometimes in Scotland. Even Ireland now complains of a 'muntjac problem', with around

300 wild sightings since 2006. And whenever the deer finds
suitable habitat – which includes not just woodland but large
suburban gardens, cemeteries, allotments and parks – its population soars.

What is the secret of its success? A big part of it seems to
be an unusual capacity to reproduce all year round. Whereas
other deer in Britain reproduce once a year, the lot of the
muntjac doe, which can reach sexual maturity as early as just
seven months, is perpetual pregnancy: within days of giving
birth, following a seven-month gestation period, she's on heat
again and ready to be 'covered' once more. In the absence of
culling, the only thing slowing muntjac right now is the fact
that females produce just a single kid at a time – versus usually
two, in the case of roe; and if the muntjac youngster is born
during a hard winter and doesn't find shelter, the chances of
survival are reduced.

The Chinese water deer is another notable Woburn fugitive.
This is a small and dainty animal, whose males lack antlers,
a deficit compensated for by tusk-like upper canines which
earns the species its nickname, the 'vampire deer'. The first
of several specimens were brought by Herbrand Russell to
Woburn in 1896, and in 1913 the count stood at 126. By the
middle of the twentieth century, escapees were living wild in
southern England. As is often the case with non-natives on
the up in Britain, the Chinese water deer seems to have
declined in its home range, although numbers persist in the
Yangtze Basin and are beginning to build in the Koreas. The
British contingent – today put at around 3,600, possibly more
– represents a significant proportion of the deer's worldwide
population. Often to be seen lounging about in arable fields
across Bedfordshire, Cambridgeshire and East Anglia (although
reaching their highest densities in wetlands), Chinese water

deer are however thought unlikely to match the success of the longer-lived and tougher muntjac.

If anything can offset Herbrand's association with the muntjac, it would surely be his pivotal role in saving another Asian deer, the Père David's. A curiosity to see this large and ungainly animal with back-to-front antlers had prompted me to travel southwest from Thetford Forest to Woburn, a 120-kilometre journey in which the austere flatlands of East Anglia gradually softened into the plush countryside of South Bedfordshire. As I rumbled over the cattle grid at the entrance to Woburn's famous deer park, a sign warned me of '*Horse-drawn vehicles and animals*', while another kept a running total of deer killed on the road – eight so far that year.

Arriving at the estate offices, a cluster of low-rise red-brick buildings sited a respectful distance from the abbey itself, I was met by Martin Harwood, Woburn's fresh-faced and newly appointed deer manager. During our subsequent tour of the well-tended parkland, Martin's passion shone through. 'I got into deer at a young age, and it's a bit of a bug, to be honest. Once you start reading about them and learning about them, you just get sucked into it.' He enthusiastically pointed out fallow and axis, sika and red. We even spotted a muntjac loitering in a patch of woodland. '*Beautiful*. Look at that shiny coat,' he marvelled, thoughtfully killing the engine of his 4x4 while I snapped away with my phone.

Martin and I soon encountered a dozen or so specimens of Père David's feeding under some copper beeches. The large, woolly-coated animals were splattered in mud, testament to their penchant for Woburn's many streams and pools; even in the dead of winter they would break the ice to stand in a pond. In China they were known as *milu* or *sibuxiang*, the latter roughly translating as 'none of the four alike', which is a reference to its designed-by-committee bodyplan. 'They say they've got the feet of a cow, the tail of a donkey, the head of a horse and the antlers of a stag,' explained Martin. 'Fat as butter at the minute,' he added, 'they are looking well. If you came back in six weeks' time, they'd all be skinny as a rake after the rut. Use up a lot of energy during the rut.'

Woburn's Père David's contingent currently stands at around 450 animals; indeed, Martin and his colleagues are forced to cull a proportion annually to prevent overcrowding. Yet not that long ago the species, which once ranged across China, Japan and Korea, had pretty much disappeared in the wild due to uncontrolled hunting. A single captive herd – recently shown to have been related to animals from the south Chinese island of Hainan – remained under strict guard at Nan-Hai-Tze, the imperial walled hunting park close to Beijing. During the mid-1860s, the French missionary and naturalist Père (Father) Jean Pierre Armand David got wind of the emperor's deer – then numbering just 34 individuals – and bribed a guard to sneak out a pair of antlers and two skins which he shipped to Paris. Père David thought them reindeer, but the animals proved new to science, and in 1866 were named in his honour. (During his travels in northern and western China, Père David discovered hundreds of other plants and animals unknown to Europeans including the giant panda and the Mongolian gerbil. As well as the deer, the wandering priest has lent his name to a vole, a mole, a snake, several birds and plants including the butterfly bush *Buddleia davidii*.) Over the ensuing decades, several live milu were smuggled to zoos and parks in France, Germany and Britain, including Woburn.

All this thievery proved to be the deer's salvation: in 1894, the emperor's herd was set upon by hungry peasants after a flooding river washed away a section of the park's wall, and the rest were killed by soldiers during the 1899–1901 Boxer Rebellion. Meanwhile, 8,000 kilometres to the east, the 11th Duke of Bedford had grown attached to this unusual deer and by 1903 had gathered together at Woburn a herd of some 18 specimens. Through careful husbandry, he got 11 of them breeding, setting the deer on the slow road to recovery. During the 1980s, Robin Russell, the 14th Duke of Bedford, and his son Andrew (now the current duke) oversaw the successful reintroduction of the species to its homeland, with a total of 38 deer from the Woburn herd released into the Nanhaizi Milu Park, the site of the former Imperial Game Park where Père David himself first observed them. Today, the global tally of milu is upwards of 5,000, encompassing several free-living populations across China, and every last animal is descended from Woburn stock. So far, the Père David's doesn't seem to have plodded into our countryside like its compatriots, the muntjac and Chinese water deer – its size counts against it – but maybe one day it might.

Herbrand Russell didn't confine himself to deer. One early visitor to Woburn described his bewilderment at 'the simultaneous sight of a dozen strange beasts feeding together in a kind of Garden of Eden'. Mouflon, black-and-white yaks, and tahrs – a form of Asian wild goat – could be seen sheltering under chestnut trees, while zebra and their foals grazed near the drive. Fenced enclosures, meanwhile, contained various endangered creatures, including Soay sheep from St Kilda island, Przewalski's wild horses from Mongolia and European bison, along with cavies, kangaroos and wallabies. Perhaps inspired by his mentor Frank Buckland, the duke even attempted to domesticate eland, castrating the bull calves and rearing them for the table, but pronounced the meat inferior to that of deer, cow or sheep.

The bird collection was impressive too, with emus, storks, cranes, Australian brush turkeys, rare geese and other water-fowl roaming free, their calls and shrieks delighting Herbrand and Mary as the pair took evening strolls in the grounds. The Bedfords majored on pheasants, including the Reeves' pheasant – another species collected in China by a certain tea inspector (or his son). Early trials weren't successful, with the duchess telling one newspaper in 1899 that 'anyone who tries the experiment will soon learn that he has done the local bird-stuffer a good turn'. A few years later, however, flamboyant males were to be admired wandering Woburn's woodland with metre-long tails and 'black and gold coats shining like a mandarin's jacket'. There's also a story from the 1950s of a Reeves' pheasant flying through the window of an office in the Dunstable gasworks only to be eaten by the employees. But the pheasant, unlike its muntjac namesake, failed to estab-lish in the wild.

The Lady Amherst's pheasant, another handsome bird from the Far East, didn't fare much better. Named after the wife of the Governor General of Bengal, a specimen first arrived in England in 1828, with the species introduced at Woburn during the 1890s. By the early 1970s, as many as 200 pairs were rooting about in the surrounding fields and woodlands of Bedfordshire, Buckinghamshire and Hertfordshire, but they've pretty much died out since. In his book, *Mrs Moreau's Warbler*, the British nature writer and ornithologist Stephen Moss describes a 1987 expedition to Ampthill (eight kilometres from Woburn) to spot the fast-vanishing avians: 'Having drawn a blank, I returned to where I had parked my car, only to notice a group of birds at the back of a field. Four magnificent Lady Amherst's pheasants, their impossibly long tails barred with black-and-silver, were feeding unobtrusively along the end of a wood. It was the only time I ever saw the species in Britain.' In a twist of fate, the Reeves' muntjac's proclivity for understorey browsing may have caused the demise of the Lady Amherst's pheasant. The latter relies on dense vegetation to avoid predators and so as the deer population increased around Woburn, the ostentatiously decorated pheasants were left exposed to foxes and magpies. A case of what Woburn giveth, Woburn hath taken away.

Perhaps the most beautiful of all non-natives, bird or otherwise, to establish in Britain, thanks in part to the Woburn experiment, is the mandarin duck. The exquisitely vivid markings sported by the males of this tree-nester from Siberia, China and Japan have captivated European travellers to the Orient for centuries. It continues to delight: when in 2018 a single mandarin drake showed up in New York City's Central Park, having escaped a private collection nearby, it triggered a global media frenzy. The species was brought to England as early as 1745 by Sir Matthew Decker, director of the East India Company. It was not bred here until the 1830s, when the Zoological Society of London purchased two pairs for Regent's Park zoo, from the Earl of Derby's Knowsley waterfowl collection. Then, in 1904, the Duke of Bedford acquired mandarin ducks for Woburn, whose many ancient trees offered abundant nesting holes. Benefiting from a steady supply of acorns, sweet chestnuts and beechmast, and from the strict control of predators across the estate, the population soared to 300 within ten years of their introduction. The duck has since colonised much of southeast England, although, again, the present-day distribution results from multiple separate translocations rather than natural spread from Woburn.

As with the Chinese water deer, the mandarin duck seems to be doing a lot better here than in its Asian homeland.

Herbrand even stocked his lakes with introduced fish, but in this case, at least, he was following his father's lead. In 1878, Francis had taken delivery of two dozen large zander, also known as pikeperch, a bug-eyed predator with vampiric teeth, long coveted by Buckland and his fellow acclimatisers. The fish, each weighing about a kilo, were sourced from a Mr Dallmer, chief fishing master of Schleswig-Holstein in northern Germany, and shipped from Hamburg aboard the *Capella*. Met in London by the duke's servants, the arrivals were escorted on to Woburn via train then horse-drawn carriage, and placed in a nine-hectare lake, described as 'full of small fish but no pike, the gravelly bottom being eminently suitable for Zander'.

Two years later, Francis also acquired 70 specimens of wels, a monstrous variety of eastern European catfish that grew several metres in length. Previous efforts had been made to introduce the species, including in 1853 at Morton Hall in Norfolk, by the British and Ottoman adventurer Sir Stephen Bartlett Lakeman, following a fisheries expedition to the Hungarian Danube. Within a decade, wels were reputedly living wild in the nearby River Wensum having slithered to freedom through a broken grating, although it's unclear whether they survived very long. In 1864, *The Field* magazine sponsored another attempt, with 14 specimens, installed in Francis Buckland's pond at Twickenham, *The Times* hailing it as the most important introduction since the turkey.

At Woburn, both the zander and wels survived, if not thrived, with Herbrand introducing further batches. From the early twentieth century, the two fish were translocated from Woburn – and from various other collections – to numerous rivers, lakes and ponds across southern England for the benefit of anglers. It is still permitted to introduce wels into enclosed waters, but the zander, which has since colonised the Severn, Trent and Thames river systems, is now *persona non grata* because its ravenous feeding habits can decimate populations of smaller fish. Elsewhere however, for instance, in parts of Scandinavia, the species remains highly prized for the pot, and in Denmark even enjoys legal protection.

There's one final introduction for which we ought to thank the 11th Duke of Bedford. An animal that has spent a century or more spreading inexorably north in a grey tide, accused of devastating forests and snuffing out its gentler, cuter, home-grown counterpart; a rodent that bosses suburban bird-feeders from Brighton to Aberdeen; a bark-stripper, a cable-chewer, a fire-starter, a plague-spreader, a thief, a vandal; a bushy-tailed, over-sexed killer with a weakness for Nutella; a species exciting such animosity that military-grade technology is ranged against it. We speak, of course, of the grey squirrel or, more precisely, the *eastern* grey squirrel; a more retiring version native to the west coast of the United States and Mexico has stayed put.

How times have changed. Back in the nineteenth century, grey squirrels were all the rage among the British upper class. There are rumours of earlier releases in Wales and Kent, but Thomas Brocklehurst, a Cheshire-based banker and silk manu-facturer, is often credited with the first confirmed introduction. Returning from a business trip in 1876, the globetrotting textiles magnate was said to have installed a pair of grey squirrels at Henbury Park, his 240-hectare estate outside Macclesfield. Whether or not Brocklehurst's animals survived is academic, since many others of his ilk were soon getting in on the act. And none with more zeal than Herbrand Russell.

You won't find 'Squirrel Spreader-in-Chief' amid his undoubtedly voluminous entry in *Debrett's*, yet the duke has one of the better claims for the title. A good proportion of Britain's current 2.5-million-strong grey squirrel population is thought to descend from ten animals introduced from New Jersey to Woburn in the 1890s. These multiplied and dispersed into surrounding woodland. Despite their reputation, recent research shows greys aren't actually good at spreading themselves. But no matter, Herbrand saw to it that deliberate releases of Woburn squirrels ensued at estates up and down the country, as well as in Regent's Park in London. With an aristocratic tailwind, augmented by further introductions from North America and liberated from natural predators, such as raccoons, skunks and large forest hawks, there would be no stopping the species. The last significant influx of grey squirrels to the British Isles occurred on the eve of World War I (and importation was banned in 1938). Since then, repeated internal transplantations, intentional or otherwise, have continued much to the disgust of red squirrel lovers, who, as we see later in these pages, are fighting a determined rearguard action against the grey menace.

Herbrand's influence on zoology extended far beyond Woburn. In 1899, he was elected president of the Zoological Society of London, a post he would hold for almost 40 years overseeing, and helping to finance, the construction of an aquarium at Regent's Park zoo. His scientific contribution was recognised by his 1908 appointment as a Fellow of the Royal Society. Even in his autumn years, the duke's passion for acclimatisation was undimmed. In 1931, a new zoological park was opened at nearby Whipsnade, to which he presented, along with the Indian muntjac, 32 Chinese water deer, a pair of Reeves' muntjac and three turkeys. With these and other animals allowed to wander freely with minimal interference, Whipsnade would be a natural extension of the Woburn project.

Herbrand's final years seem to have been melancholy. He was alienated from his only son, Hastings, a pacifist who had refused to fight in World War I. Then, he lost his wife in an air accident. Mary had completed record-breaking flights (with a co-pilot) to India and South Africa in the 1920s, and at the tender age of 68 had gained her own licence, earning the nickname 'the Flying Duchess'. But during a solo flight in 1937, she disappeared in poor weather off the coast of Norfolk. With Herbrand's own death three years later, the grand Woburn acclimatisation experiment was over, although its spirit would resurface 30 years later, when his grandson, Ian Russell, the 13th Duke of Bedford, opened a successful safari park on its grounds. And, of course, Herbrand's cherished deer park, with its spectacular herds of non-natives, can also still be visited to this day.

6

The Plant Hunters

'I thought I could do my own small part to save the planet by becoming a vegetarian. Actually, I did it not so much because I love animals but because I hate plants.'

The Big Picture: An American Commentary,
Whitney Brown, 1991

'People don't realise you get this kind of habitat in Britain. We could be in Brazil right now,' said Jake Chant, folding his gangly frame under a hazel branch hanging low over the water. I could see his point; as we picked our way down the narrow, overgrown stream abuzz with insects, its surface glinting in the late morning sunlight, we were like explorers of a tropical world hunting some exotic monster. My companion, a field officer at the Devon Wildlife Trust, even brandished a metre-long machete, although this was used as much for testing water depth and negotiating the strands of barbed-wire which occasionally blocked our way, as for hacking at stubborn vegetation. In one respect at least, the analogy held true. We *were* on the lookout for an exotic species. But this was no monster: it was a flowering plant, and a rather beautiful, fragrant one at that.

Observing the tacit rule of botanical nomenclature that every plant should boast at least half a dozen common names, our quarry – the Himalayan, or Indian, balsam – was also variously known as Policeman's Helmet, Bee-bums and Poor Man's Orchid. Further sobriquets included Jumping Jacks and Stinky Pops, referring to the facility with which its seed pods, when ripe, detonated at the slightest disturbance – be it a drop of rain or the flick of a child's finger – spraying their contents far and wide. Invariably ranking high on lists of Europe's worst invasive plants, the Himalayan balsam (a relative of the Busy Lizzie) could be seen as the literal embodiment of one of Charles Elton's 'ecological explosions'.

The species was first brought to Britain from its native foothills in Nepal, India and Pakistan in the late 1830s by John Forbes Royle. An army surgeon with the East India Company, Royle was also superintendent at the Company's botanical gardens in Saharanpur, 200 kilometres north of Delhi. During his posting, he was tasked with collecting plants promising medicinal value but, in the case of the Himalayan balsam, the beauty of its sweet-scented pink-and-white flowers was alone enough to beguile the doctor and earn a passage to Blighty. As with so many other significant plant introductions, staff at the Royal Botanic Gardens at Kew played a critical role in the early years, tenderly cultivating in greenhouse conditions what was assumed to be a delicate species. The plant needed no such cosseting: in 1855, within 20 years of arriving it was growing wild in Hertfordshire and Middlesex and, by 1890, had been declared a weed. By then, Himalayan balsam – which, when full-grown, attains heights of two metres or more, making it Britain's tallest annual plant – was spreading at an estimated rate of 645 square kilometres every year. As if its ballistic method of seed propagation wasn't enough, the balsam was also helped along by the nation's beekeepers, appreciative of its lengthy flowering season and abundance of nectar, who deliberately planted it close to their apiaries.

Today, few corners of Britain are balsam-free, the sole check on its advance being the plant's preference for the banks of rivers and streams. 'It famously likes wet feet,' said Jake, as he slashed at some stinging nettles. 'Although, who's to know what will happen over time? I've seen them out in a country lane.' In any case, the plant is already causing big changes.

Shooting up on prime waterfront early in the growing season, the balsam is accused of shading out perennial natives, such as marsh woundwort, purple loosestrife and great willow-herb, while monopolising insect pollinators with its irresistible summer-long bonanza of nectar. Then, at the end of the season, being an annual, it promptly dies off, leaving behind an empty bank which, in the absence of the stabilising roots of the displaced perennials, washes away in the first major flood of the winter, the released silts choking fish eggs and aquatic invertebrates. If all this wasn't bad enough, the Himalayan balsam may also deplete mycorrhizal soil fungi, preventing the regrowth of native plants. In the meantime, its peppercorn-like seeds – each plant fires off about 800 of them – float downstream, to be deposited on a new bank, often where an eddy forms, and the destructive cycle resumes the following year.

The favoured means of tackling Himalayan balsam is to rip it up by the roots before the trigger-happy seed pods get going. But it's a labour-intensive process, requiring multiple return visits to the same stretch of water. In 2003, the Environment Agency calculated that total eradication from the UK could cost as much as £300 million, and the government is so far relying on teams of volunteers to do the job. It was at the invitation of one such group, the Tale Valley Trust in East Devon, that I had gone in for a spot of 'balsam-pulling' myself, rendezvousing early that August morning at the elegant nineteenth-century mansion house of Escot Park. This was home to natural history enthusiast and the Trust's prime mover, Mish Kennaway – or, to give him his full title, Sir John-Michael Kennaway, 6th Baronet. In 2004, offended by the epidemic of Himalayan balsam along the Tale, a pictur-esque tributary of the River Otter which flows through the estate, Mish had opted for a zero-tolerance approach, coor-dinating a decade-long campaign whose precision and dedication would have impressed his military forebears.

The results were undeniable: whereas in previous summers, balsam plants would crowd the river bank in their thousands, today the pink-and-white menace was virtually absent along a 12-kilometre stretch of the Tale, from headwaters at Broadhembury to its confluence with the Otter. Nevertheless, insurgents remained, each capable of undoing everything with a single pop of a seed pod. So, the work went on.

Mish had rallied seven of us to his cause, including his 23-year-old daughter Olivia. We were issued waders, divvied up into pairs and deployed by Land Rover at strategic points along the length of the Tale. Mish's instructions were simple: 'Make sure you pull it up by the roots. Don't snap it off, then chuck it as far away as possible from the river.'

Over the ensuing three hours, Jake and I had moved steadily downstream encountering no more than a couple of dozen balsam which were summarily dealt with. The plants offered little resistance, the main challenge being the intact extrication of their lengthy, trailing and fragile stems. Dragonflies zipped past. At one point, a trio of bullocks on a high bank peered down at us in apparent puzzlement. Over sandwiches back at Escot, we compared results: the balsam had been cleared in record time this morning. Surely at this rate, I thought, the problem would soon be licked? Then came a cold dose of reality.

Before I left, Mish took me to the lowest reach of the Tale to deal with a few stragglers. The flow here was deep and fast, and pressed against my chest, threatening my balance. I peered under a low bridge. Beyond, where the Tale emptied into the Otter, the river banks were festooned in balsam. Tens of thousands of delightful pink-and-white flowers stretched to the horizon. One battle may have been won, but it seemed the war had barely begun. A chill seeped around my toes. My waders had sprung a leak.

'Once the sources of our essential needs are secure and we have erected a familiar, i.e., comforting, environment, we next seek to embellish this environment with diversity, including diversity in plants.' So wrote, in 2001, the invasion biologist Richard N Mack, of Washington State University, who argued that an instinct to embroider our surroundings explained why people have collected and deliberately transported plants around the world for so long. Mack was surely right.

For millennia, while plants have been chosen, grown and traded primarily as food and fuel, medicine and material, their beauty, scent, taste and all-round exoticism have charmed us too. Perhaps the world's oldest garden was created by the Chinese emperor Shennong. According to legend, Shennong – who also gets the credit for inventing acupuncture – tested countless plants for their therapeutic properties, personally tasting up to 70 different types on a single day. Alas, he perished from a ruptured intestine in 2697 BCE, having eaten the wrong kind of grass. The ancient Greeks were among the first Western civilisations to have cultivated non-native species, with the Minoans of Crete (2100–1600 BCE) sourcing date-palms and papyrus from their Egyptian neighbours.

The Egyptians themselves imported poppies from the eastern Mediterranean and pomegranates from the Caspian region. In about 1470 BCE, Queen Hatshepsut of Egypt led a sea voyage to the Horn of Africa. Reliefs in the Temple in Thebes depict her bringing back 31 small trees, thought to have been live myrrh or frankincense saplings, for planting as a source of aromatic resins. Later, the pharaoh Rameses III decorated his new town in the Nile delta with 'great vineyards; walks shaded by all kinds of fruit trees laden with their fruit; a sacred way, splendid with flowers from all countries, with lotus and papyrus, countless as the sand'. The Babylonians of ancient Iraq were famed for their – potentially mythical – hanging gardens, with trees, shrubs and trailing vines from foreign lands supplementing native tamarisk and date-palm; water was pumped, ingeniously, from the nearby Euphrates river to irrigate the plants, which bedecked a series of raised terraces. During Alexander the Great's fourth-century-BCE campaigns in India, his botanists collected seeds of ivy, bamboo, banyan, banana and many other exotic trees and flowers.

Oriental treasures had long held a fascination for Westerners, and these included horticultural ones, from roses to violets, peonies to peaches, which made their way to Mediterranean gardens via the caravans and dhows of the Silk Road. But from the late ninth century CE, as a new age of empire-building dawned, exotic plant species started flooding westward as never before. The collection and cultivation of rare botanical specimens was initially the preserve of religious orders and the landed gentry, but as prosperous new social classes emerged across Europe with the time and money to indulge in ornamental gardening, demand rocketed for new flowers, shrubs and trees. The British gardener John Tradescant the Elder (c. 1570–1638) was among those to capitalise, collecting seeds and bulbs across Europe, North Africa and the Levant for a succession of aristocratic clients keen to show off ever more extravagant gardens. In 1630, Charles I made him Keeper of his Majesty's Gardens, Vines and Silkworms. His son continued the tradition, creating a garden at Lambeth said to contain every kind of northern European plant, as well as varieties from the colony of Virginia in North America where he'd spent his early life. Other curiosities were hoarded at Lambeth, including whale ribs, seahorses, crocodile eggs and the hand of a mermaid; the collection – dubbed the 'Tradescant Ark' – formed Britain's first public museum.

Expert plant hunters from the Netherlands, France and Britain meanwhile fanned out across new colonies in Africa, Asia, the Americas and Australasia, commissioned by their own governments or self-funded, in search of the latest botanical beauty or wonder crop. From the start of the seventeenth century, naturalists in the VOC (Verenigde Oost-Indische Compagnie) – the Dutch East India Company – began collecting new plant specimens from southern Africa, the

Indian subcontinent and the Far East. Although such missions, originally at the request of the Leiden Botanical Garden, were tasked with finding 'useful' varieties, their protagonists sometimes had other ideas. Hendrik Adriaan van Reede tot Drakenstein was an amateur botanist and senior official in the VOC, who collected and described 740 different kinds of plant in the jungles of Malabar, southwest India. He was as much captivated by the beauty of these forests as by the functional properties of their constituent species. Plants weren't just gathered from the colonies, they were spread there too. In the 1650s, the VOC established an outpost on the Cape of Good Hope, in southern Africa, as a way station for ships moving between Europe and the Far East. The Dutch settlers on the Cape's barren flats set about introducing new plants as food, fuel, timber or to stabilise the shifting sands. These experiments in translocation were supported by the establishment in 1660 of a botanical garden, the forerunner of Cape Town's Kirstenbosch Gardens. Today, almost 9,000 types of non-native tree, shrub and flowering plant grow wild in the region, of which 161 are classed 'invasive', although the Cape retains an extraordinary number of endemic species.

Founded around the same time as the VOC, the British East India Company initially enjoyed far less trade than its Dutch counterpart, and struggled to compete with the Portuguese and Spanish. It nevertheless steadily extended activities across the Middle East and Asia, building up a trade in cotton, silk, indigo, saltpetre and slaves, and blossomed from the late eighteenth century following the 1757 conquest of Bengal in India. British colonial botanising got started in a big way, ostensibly to feed the native subjects of the rapidly expanding empire then being hit by wave after wave of famine. Botanical gardens were established in Calcutta, Madras and Bombay to cultivate imported food crops, such as sago palm from Malaya. Similar botanical stations were founded elsewhere, including on the West Indian island of St Vincent in 1765; a few decades later Captain Bligh, having survived the mutiny on the *Bounty*, brought breadfruit seedlings here from Tahiti as food for the slaves. But any humanitarian motivations for botanical collection among the British were superseded early on by a more powerful instinct to compete in the highly profitable trade in plant-based commodities. And one man would do more than most to pioneer global plant-hunting.

Joseph Banks was a gentleman naturalist who had risen to scientific prominence after joining James Cook's first voyage to the South Seas aboard the *Endeavour* in 1768. During the three-year expedition, Banks and his assistant Daniel Solender collected some 3,000 different plant species, a third of them new to science. In 1773, Banks was informally appointed director of the Royal Botanic Gardens at Kew by his close associate George III. He quickly transformed the king's modest pleasure garden into the nerve centre of an international network of botanical gardens taking in stations in the West Indies, Cape Town, India, Sri Lanka, Malaysia, Singapore, Hong Kong, Australia and New Zealand. Coordinated from Kew by Banks and his successors, notably Joseph Hooker, these outposts served as testing beds for cinnamon, coffee, black pepper, tea, sugar, tobacco, indigo, opium and other prized crops, access to which was fiercely guarded by Dutch, French, Spanish and Portuguese rivals. The worldwide movement of live botanical specimens was facilitated by the invention in the 1830s of the Wardian case. The brainchild of a physician, Nathaniel Bagshaw Ward, this was, in effect, a portable greenhouse that could be sat on a ship's deck, protecting plants from salt-spray in their own humid microclimate. The innovation allowed live seedlings to be transported far longer distances than ever before.

Plant hunters, both professional and amateur, were commissioned to discover and collect valuable new species and varieties. Often the plants had already been 'discovered', and it wasn't so much 'collecting' as stealing. For instance, in 1848, on behalf of the Royal Horticultural Society and British East India Company, the Scottish plant hunter Robert Fortune used Wardian cases to sneak 20,000 tea seedlings out of China. The booty was replanted in Darjeeling, India, thus breaking the Chinese tea monopoly. Similarly, in 1860, Kew and the India Office dispatched agents to Peru, Bolivia and Ecuador to smuggle out seeds and live specimens of cinchona under the nose of the authorities. Also known as the 'Andean fever tree', cinchona was the source of quinine, a bark extract, then the best-known treatment for malaria. The mission was successful, with the species translocated to India, where colonials and indigenous people alike were being ravaged by the disease. Eventually eclipsed by a more effective strain cultivated by the Dutch in Java (also from smuggled seed), Britain's early cinchona plantations nevertheless helped bolster an expanding empire. A few years later, Henry Alexander Wickham pilfered 70,000 seeds of the rubber tree from the Brazilian Amazon allowing the British — again via Kew – to establish successful rubber plantations in Sri Lanka and Malaysia.

Back in Britain, growing interest in ornamental gardening was fostering a new industry. In 1808, John Veitch of Exeter founded what would become the country's largest group of commercial family-run nurseries. Just like Kew, the Veitches and their ilk commissioned overseas plant collectors and acted as a hub for international exchange; fashionable new varieties sourced by British horticulturalists quickly popped up across

Europe and North America. By 1914, the Veitches alone had introduced some 1,300 new species into cultivation, although few naturalised. Indeed, commercial nurseries and botanical gardens can't be directly blamed for many escapes in Britain. Kew, for instance, despite housing some 10 per cent of the world's known varieties of plant, is directly responsible for a mere handful of introductions in this country, notably the gallant-soldier, a weed in the daisy family brought from Peru in 1796 and recorded growing wild by 1863. A far greater number of non-native ornamental plants made their bid for freedom from private gardens, both big and small, and especially as a result of a new trend: wild gardening.

This late-Victorian mania was championed by an Irish journalist called William Robinson, who objected to the formal, classical-style landscape design of the time, with its statues, water features, fussy topiary and over-reliance on delicate, short-lived annuals. In *The Wild Garden* (1870) Robinson called for the 'naturalizing or making wild innumerable beautiful natives of many regions of the earth in our woods, wild and semi-wild places, rougher parts of pleasure grounds, etc. and in unoccupied places in almost every kind of garden'. Fast-spreading, hardy exotics, able to thrive in a temperate climate, such as rhododendrons, Himalayan balsam, giant hogweed, buddleia and Japanese knotweed, were by then cultivated in Britain, but Robinson's writings boosted their popularity, setting them on a path to invasiveness. During the late nineteenth and early twentieth centuries, foresters and estate managers caught the wild gardening bug too, shifting from traditional coppicing of indigenous species to the introduction of non-native trees and shrubs, often as cover for game. But social changes in the aftermath of World War I resulted in the neglect of many great country estates. The environmental geographer Ian Rotherham of Sheffield Hallam University writes: 'With the controlling hands of thousands of gardeners now removed . . . a host of exotic and often aggressive plants leapt the garden wall to freedom.' Decades later, ecologists twigged what had happened. By then, as Rotherham puts it, the 'genie was out of the bottle'.

In the late afternoon of Thursday 14 September 2015, two exhausted tourists clambered aboard a lifeboat on Lough Leane, Killarney National Park, southwest Ireland. They'd been guided to the lakeside by rescuers in a coastguard helicopter, after an air, land and water rescue operation launched three hours earlier when the pair lost themselves in a dense purple forest of rhododendron. The non-native shrub, covering a third of the 10,000-hectare park, had long dismayed conservationists and politicians. Since the 1980s, volunteers and the park authorities had been struggling to control the plants swamping Killarney's ancient oak and yew woodland. A small fortune had been spent on herbicides and there were calls for the army to be brought in. The incident wasn't the first of its kind. just the previous year two experienced hill walkers had had to be led to safety after getting into difficulties in rhododendrons infesting the nearby Knockmealdown mountains. A rescuer who'd spent hours crawling through the tangled vegetation – some of it spanning 15-metre drops – described the operation as one of the most dangerous of his career.

First brought as ornamentals to Britain by the commercial nurseryman Conrad Loddiges in 1763, rhododendrons are among the earliest non-native shrubs to be planted in large numbers. The most widespread variety in Britain – one with spectacular and profuse pink and purple flowers – is actually a hybrid. It was created by interbreeding a Spanish subspecies of the pontic rhododendron (also native to the Caucasus region), with two American types. Intended as a basic grafting stock for other rhododendron varieties, the crossing proved to be a Frankenstein's monster.

The new rhododendron thrived in the British climate, growing up to eight metres in height. Victorian wild gardeners loved it, and so did gamekeepers, who planted the evergreen shrubs as shelter for pheasants. But with each highly vigorous plant producing a million dust-like seeds, by the 1890s the rhododendron was out of control, especially in southwest England, Wales and western Scotland, where it formed dense monocultures putting everything else in the shade. Unlike many of its competitors, the plant was adept at recolonising after fires, including those deliberately set by gamekeepers. The shrub eventually reached almost every county of Ireland, as well as other parts of northern and eastern Europe, its advance checked only by a requirement for acid soils. It even invaded New Zealand after establishing there in the 1950s.

Rhododendron control is a challenge, to say the least. As with Himalayan balsam, parties of machete-wielding volunteers have been the first resort, especially on less accessible terrain where mechanical removal is not an option. Mature plants are prioritised due to their seed-dispersing potential, but any new growth must also be tackled. Lundy island in the Bristol Channel has seen a decades-long campaign against rhododendrons, which were introduced to the island's Millcombe Valley in the nineteenth century and proliferated following a brush fire in 1926. Since then, extensive stands threatening the island's unique Lundy cabbage – and, in turn, two endemic species associated with it (a beetle and a weevil) – have been cut and burnt, and any new growth clobbered with herbicide. In 1997 alone, volunteers spent 226 hours to clear a single hectare. Twenty years on, the battle is almost won with Dean Woodfin Jones, the island's current warden, recently reporting that there was now 'next to no new rhododendron flowering'.

The rhododendron's rap sheet is a long one. Perhaps, its main crime is ousting native woodland plants, which would otherwise support a greater range of wildlife. Few, if any, British insects make a home among rhododendrons. The number of earthworms, birds and hazel dormice also slumps in infested forests, which become an ecological desert. The leaves are poisonous, especially to grazing animals, earning rhododendrons the wrath of livestock farmers. Even those pretty pink and purple blooms have a sinister side: the honey derived from rhododendron nectar contains hallucinogenic neurotoxins implicated in so-called 'mad honey disease'. This condition was described as early as 401 BCE by the Greek philosopher-warrior Xenophon, who reported honey-gorged soldiers in the rhododendron fields of Turkey suffering from vomiting and diarrhoea, acting crazy and lying down in a stupor. The next day the befuddled troops awoke as if from a drug-fuelled bender.

Rhododendrons are also suspected of allelopathy – secreting compounds from their roots to prevent nearby plants germinating and growing – but there's little evidence for this; the year-round gloom cast on the forest floor by a dense canopy of mature rhododendron is probably enough to snuff out any competition without resorting to chemical warfare. Rhododendrons are, however, known to host *Phytophthora ramorum*, a fungal-like pathogen causing extensive, and sometimes fatal, bleeding cankers on the trunks of a wide variety of trees and shrubs. (*Phytophthora* translates as 'plant destroyer'.) First described in North America in the mid-1990s, 'sudden oak death', as the disease is sometimes known, arrived in the UK in 2002 where it mainly infects larch, although beech is susceptible, as are non-native red, holm and Turkey oaks, along with sweet chestnut and horse-chestnut. As bad as rhododendrons seem to be, another nineteenth-century introduction from the Caucasus would generate far more concern.

Lacking the gaudy attractiveness of Himalayan balsam and rhododendron, its dimensions were largely what recommended the giant hogweed to the Victorians. With all the appearance of cow parsley on steroids, Europe's tallest herbaceous plant was extolled as 'magnificent' by the pioneering Scottish landscape architect John Loudon, while William Robinson was enamoured of its 'bold foliage'. Others described its proportions as 'Herculean', and its 'splendid invasiveness' was recognised early on – as a *good* thing. Giant hogweed was even grown in Buckingham Palace Gardens, while up at Woburn, Herbrand Russell pronounced it his favourite plant. Elsewhere in Europe giant hogweed and the related Sosnowskyi's and Persian hogweeds were cultivated as bee pasture and silage; cows, sheep and pigs apparently enjoyed eating the stuff. Norwegians affectionately knew it as 'Tromsø palm'. More than a century would pass before anyone seemed to notice a downside.

Then, during the 1970s, children in Britain began showing up in hospital with blisters on their hands, and around the mouth and eyes. The kids had been fashioning telescopes and blowpipes from giant hogweed stems which, on investigation, were revealed to contain chemical irritants called furocoumarins in the sap and bristles. These light-sensitive compounds, also present in a variety of hogweed relatives such as celery, carrot and parsnips, as well as in citrus plants, are a natural defence against herbivorous insects, but when in contact with human skin, they react under ultraviolet radiation to cause a burning sensation, and sometimes serious and long-lasting rashes. The giant hogweed's stock plummeted overnight as these once-admired curiosities were recast in the press as real-life triffids, an allusion to John Wyndham's bestselling science fiction novel *The Day of the Triffids*, in which humans, blinded by a meteor shower, are picked off by monster plants. The Wildlife and Countryside Act 1981 designated the giant hogweed as a noxious weed, making it an offence to plant or cause it to grow in the wild, punishable by a fine of up to £40,000 or imprisonment. By then it was far too late.

Giant hogweed had long been a fixture in the British countryside, having naturalised just 11 years after its first appearance at Kew in 1817. Like Himalayan balsam, the giant hogweed spreads its seeds – 20,000 per mature plant – along watercourses, as well as on the wind, and it now colonises the banks of many major rivers such as the Tweed, Clyde, Usk and Wye, where if untreated, hogweed forests can develop up to five metres in height. Given its toxic sap, giant hogweed removal doesn't lend itself to teams of enthusiastic volunteers. Instead, landowners generally have to pay herbicide-wielding specialists in protective clothing to do the job, although in Aberdeenshire, trials have been underway since 2013 using sheep to graze the hogweed; for some reason, the furocoumarins don't seem to affect the animals.

Now we come to what is the most talked-about invasive plant these days: Japanese knotweed. First cultivated in Chiswick, London, in the early nineteenth century by the Royal Horticultural Society (who thought it was something else), the species is most closely associated with Philipp von Siebold, a Bavarian naturalist, who in the 1840s collected it from Japan's volcanic slopes for his nursery at Leiden in the

Netherlands. In 1850, knotweed specimens turned up at Kew and soon afterwards the plant's fast-growing bamboo-like stems, heart-shaped leaves and spikes of creamy-white blossom began to enrapture Victorian gardeners. The plants were sold in nurseries across Britain and cuttings widely shared. William Robinson, a big fan, described knotweed as 'most effective in flower in the autumn' and advised planting it in groups of two or three, although even he warned that the knotweed 'cannot be put in the garden without the fear of their overrunning other things'. That didn't worry nineteenth-century engineers who also used knotweed to stabilise railway embankments.

In this country, with the possible rare exception of a few unusual hybrids, Japanese knotweed cannot yet produce viable seeds, and so far, every single plant seems to be female. This hasn't held it back, for the species instead propagates itself asexually from the tiniest fragment of rhizome. As soon as gardeners began chucking out knotweed root material, the plants were on their way. Japanese knotweed was first recorded growing wild in 1886 at Maesteg, South Wales – where it causes problems to this day – and in 1900 had established in London. By the 1960s, the plant could be found from Cornwall to the Outer Hebrides. Today, what is effectively a single female clone occurs right across Europe, as well as in North and South America, Australia and New Zealand. The only place the species seems to struggle is back home on the lava fields of Japan.

Although Japanese knotweed thrives on river banks, alluvial forests, roadsides, refuse tips, wasteland and other disturbed habitats, its greatest economic impacts are felt in towns and cities. Back in Asia, the rhizome systems – which can grow by up to several metres each month – had evolved to squeeze between volcanic rocks, and in this country are accused of undermining foundations, blocking drains and even pushing through solid concrete. Removing plants is an expensive business, since the entire root system, which can extend for up to 20 metres underground, has to be killed, typically with repeated injections of herbicide into the stem. It cost £70 million to eliminate Japanese knotweed from the Olympic Park in east London – along with giant hogweed – in readiness for the 2012 Games, and total nationwide eradication of the species is put at £1 billion. Thus, much like dry rot, asbestos or plans for a new high-speed rail link, the discovery of a single knotweed plant on, or even near, a property sends mortgage lenders running for the hills. (A 2018 survey found that the plant has knocked £20 billion off the total value of the UK property market.)

Some experts say concerns over knotweed are wildly exaggerated, bordering on hysteria, and point out that buddleia, the much-loved 'butterfly-bush', is just as destructive. Nevertheless, in 2010 the British government approved trial releases of psyllid bugs (yes, that *is* a silent 'p'), sap-sucking insects known to attack the knotweed back in Japan. The idea is that they'll do it here too. Nearly a decade on and it's still work in progress.

Volunteers, like those of the Tale Valley Trust, are dedicated and hard-working, yet their efforts do little to check the spread of invasive plants on a national scale. Local authorities and other landowners therefore continue to rely on a small army of specialist contractors to battle the unwanted offspring of our love affair with ornamental plants. And even the paid contractors are now struggling: not only are many chemical treatments losing their potency as many weeds evolve resistance, but ever more stringent regulations threaten to curtail their use altogether.

Glyphosate, the active ingredient in 'Roundup' and many other herbicides, is a case in point. Used by farmers the world over to tackle agricultural weeds, glyphosate is also regularly deployed in this country against Japanese knotweed, Himalayan balsam, rhododendron and giant hogweed, and among the few chemicals approved for use in or near water. But, as with many pesticides, campaigners worry about glyphosate's 'non-target effects', since it will kill any plant to which it is applied. Fears are also growing that it could even harm people, particularly after a 2018 court case found Monsanto, the manufacturer of Roundup, liable for cancer developed by a California school groundskeeper who had allegedly used the product for many years. Although the European Union recently voted to renew glyphosate's licence for a further five years, pressure is growing to ban it altogether. Meanwhile, manufacturers are withdrawing from sale other, more specialist, herbicides, claiming that the rigorous – and expensive – testing procedures now required under EU legislation makes them no longer viable to produce.

With the end possibly in sight for traditional weedkillers, the hunt for alternatives is hotting up. Several weeks after my balsam-pulling outing on the Tale, I headed to Exmoor to check out a promising new weapon in the war on invasive plants. This vast tract of desolate heathland on the border of Somerset and Devon, cut through with wooded valleys and rising in the west to England's highest sea cliffs, has all the appearance of a pristine wilderness untouched by people. It's easy to see why the place was designated as a National Park in 1954. Yet, like the rest of Britain, Exmoor's water, minerals, wildlife, timber and other natural resources have been exploited for millennia, the landscape incessantly hunted, drained, grazed, ploughed and swaled (the local term for controlled burning). And, just like everywhere else, assorted unwanted non-natives, deliberately or accidentally introduced, have established here and are spreading. Among the plant invaders were many of the usual suspects – Himalayan balsam, Japanese knotweed, pontic rhododendron – along with newer arrivals, including montbretia, American skunk cabbage and Himalayan knotweed. The latter is a distant but equally intractable cousin of the Japanese variety, distinguished by narrower, spear-like leaves.

In the village of Simonsbath, pretty much in the dead centre of Exmoor, I met a pair of National Park Authority conservation officers, Ali Hawkins and Heather Harley. They'd invited Trevor Robinson, the sales manager from a start-up company called RootWave, to demonstrate his company's chemical-free solution to the problem of unwanted vegetation.

'The "RootWave Pro" basically uses electricity to boil the fluid in the plant and destroy the cell structure right down to the roots,' said Trevor, sporting sunglasses, a high-visibility jacket and yellow safety boots. Clutching a lance-like implement operated by a trigger handle, he strode over to a three-metre-high clump of Japanese knotweed, which was sprouting impudently from a roadside verge. Behind him trailed several metres of heavy-duty orange hosing hooked up, in the back of his white Ford Ranger pickup, to a nondescript grey control box. The latter was powered by a large petrol generator growling like an angry lawnmower.

'There are various voltage settings, depending on the species,' roared Trevor over the din, as he carefully positioned the tip of the lance close to the base of a single knotweed stem. 'Three-thousand is enough for moist plants like balsam or hogweed, whereas dry, woodier, stuff like this needs the full five-thousand!'

The three of us watched from a safe distance as he squeezed the trigger and a yellow spark flashed across a tiny hook at the far end of the lance. There was a sizzling noise, along with occasional bangs and pops. White steam started billowing from the knotweed thicket, and the air filled with a not unpleasant aroma, reminiscent of joss sticks. Presently, the bamboo-like stem keeled over and Trevor started working on the next one.

A quarter of an hour later, with the generator off and lance safely stowed back in the pickup, we examined the knotweed. Beyond a few blackened and broken stems, there was precious little sign of its recent ordeal. According to Trevor the real damage was done underground, although repeated treatments would almost certainly be needed, perhaps over many years, for larger, tougher, stands. (I couldn't judge whether Ali and Heather were convinced, although I later discovered that they want to purchase the technology to carry out further trials on Exmoor. RootWave Pro is also garnering interest overseas, with machines in France, Canada, South Korea and New Zealand.)

We moved to a second site, an adjacent meadow through which meandered a stretch of the River Barle. An isolated clump of knotweed had erupted from the bank, mere centimetres above the waterline. Doubtless mindful of what sort of combination 5,000 volts and a river might make, Ali expressed her concerns. 'Don't worry. Piece of cake,' smiled Trevor, for whom such challenges seemed all in a day's work.

Just how many introduced plant species have today established in Britain? As ever, estimates vary. According to Helen Roy at the Centre for Ecology and Hydrology, at least 1,506 types of non-native plant were established in England, Scotland and

Wales in 2017, of which 101 caused some form of negative impact. Far higher numbers are thrown around. What is certain though is that many, if not most, of the troublemakers were intentionally introduced via the horticulture trade. Conservation charities insist that recent European-wide restrictions on trade in a handful of the worst culprits, such as American skunk cabbage and South African curly water-weed, are inadequate. They also point to the lack of effective monitoring, with wholesalers, garden centres and pet shops sometimes selling banned species. Some are even accused of flouting the rules by renaming prohibited plants and duping their unsuspecting customers.

'Without any screening in place we're playing a game of Russian roulette with our countryside and native wildlife,' said Trevor Dines of Plantlife in 2016. The huge growth in online trade in plants exacerbates the problem. A recent study of international e-commerce found that more than 500 kinds of plant regarded as invasive were offered daily on internet auction sites such as eBay. Of these more than a dozen appear on the International Union for Conservation of Nature's list of 'World's Worst Invasive Alien Species'. With close to a half billion pounds' worth of live ornamental plants being imported to Britain every year, it's a sure bet that further invasions are on the horizon.

And, along with the plants, come stowaways.

7

Unwanted Hitch-Hikers

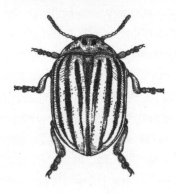

'We stop the Press, with very great regret, to announce that the POTATO MURRAIN has unequivocally declared itself in Ireland. The crops about Dublin are suddenly perishing.'

The Gardeners' Chronicle, Saturday 13 September 1845

There's a knack to finding terrestrial flatworms. For a start the weather's got to be right: you need a few days of decent rainfall to tempt them to the surface. Next, you've got to pick your habitat, preferably somewhere dark and dank. Then it's down on your knees, gently peeling apart moist leaf litter and checking under logs and large stones for clods of mucus-stuck soil, a possible sign of flatworm activity. You're looking for something no more than a few centimetres in length, possibly much smaller, a tiny glistening yellow, black or brown creature that looks like a slug except for its smooth skin and a lack of eye-stalks. (For the taxonomically minded, slugs are molluscs, whereas flatworms belong to their own group, the Platyhelminthes, which also counts among its number parasitic flukes and tapeworms.)

I learned much of this from David Fenwick, a trained horticulturalist and accomplished wildlife photographer, who spends his life documenting the fauna and flora – terrestrial, freshwater and marine – of southwest England, particularly its most recent additions. Few people are as close to the ground, so when a mysterious new species of sea slug appears on a pontoon in a Cornish harbour, Dave's often the first to notice. In recent years, he's been dogged by poor health – he suspects a mysterious tick-borne disease – yet he tries to stay positive, noting that: 'The slower you go, the more you'll see and become aware of.'

One morning in late October, after a spell of heavy rain, Dave agreed to find me some non-native flatworms. Our hunting ground was Moresk, a wooded valley on the edge of Truro in Cornwall, tucked away behind a tenpin bowling alley, multi-storey car park and council offices, each an unabashed variation on the theme of grey. The valley itself was dominated by the supporting piers of the county's longest railway viaduct, reconstructed from a Brunel original in 1904. Dave chose this place for good reason: between 1839 and 1975 it was the base of operations for Treseders Nursery, a family-run horticultural enterprise that brought many southern hemisphere plants to Britain for the first time. And, hidden amid the countless trees, shrubs and herbs from the other side of the world that passed through Moresk was a host of invertebrate stowaways.

In the early years, Treseders had been a modest operation selling vegetables, roses and common hedge shrubs such as privet, laurel and box. Things took a turn for the exotic in 1857 when the teenage John Garland Treseder, and his brothers, Thomas and Charles, emigrated to Australia. With Cornwall's tin and copper industry in steep decline, and prospects generally looking bleak, the goldfields of New South Wales had an obvious allure. In the event, the Treseder boys stuck to what they knew, setting themselves up as successful market gardeners and supplying Old World vegetables to the fledgling colony. (Thinking ahead, John had brought with him some well-wrapped seeds.) But then John switched focus.

By the late 1870s, he had become aware of the burgeoning appetite for exotic ornamental plants in Europe and North America, fuelled by the writings of William Robinson and others, and was determined to exploit it, conducting a series of plant-collecting expeditions across Australia, Tasmania and New Zealand, as well as to outlying islands. From his Sydney nursery he also began exporting millions of seeds, representing dozens of native Australasian species, to customers worldwide. During a visit to the Blue Mountains near Sydney, John encountered in deep, shady gorges what would prove to be a signature Treseder introduction.

The tree fern *Dicksonia antarctica* was a primitive and spectacular plant, whose fibrous, trunk-like, rhizomes grew ten metres in height and held aloft a canopy of delicate fronds. The species was in fact already cultivated in a few European hothouses, but John surmised its potential to thrive outdoors in the mild maritime climate of southwest England. Exploiting the tree fern's natural ability to regenerate itself after a bushfire, he dispatched some scorched rhizomes to landowners in

Cornwall to test his theory. As his great-granddaughter Suzanne Treseder writes in her intriguing 2004 family history, *A Passion for Plants*, 'The burnt fern trunks were far easier to transport than the growing specimens, and were quickly rejuvenated by being immersed in water for some days.' John's gamble paid off and by the early 1890s he was shipping hundreds of tree fern trunks to Britain and the United States. Secreted in the cavities of these hairy trunks, and among the leaf litter sometimes used for packaging, was a hidden menagerie of invertebrates.

On the death of his father in 1895 John returned to Moresk, followed a few years later by his two sons, Ira and Jack. By now, Treseders catalogues were advertising not just varieties of tree fern, but cordylines, phormium, pittosporum, acacia and eucalyptus, sourced from an extensive network of overseas trade contacts. Their hardiness having been tested and confirmed at Moresk, the plants would be dispatched by rail; one of Ira's duties was to transport consignments to Truro station by donkey. At the gentle urging of the Treseders, many a seaside resort in southwest England and Wales took on a distinctly Antipodean character. Moresk itself was festooned with subtropical foliage and, foreshadowing the later success of garden centres, the nursery and surrounding valley became a favoured visitor attraction.

'Customers were encouraged to call in and enjoy these convivial surroundings at their leisure,' notes Suzanne Treseder. 'This suggestion was clearly successful as, with the increasing popularity of the motorcar, many wealthy and horticulturally inclined clients began calling at the nursery in person to select the plants for their gardens, rather than leave the decision to their gardeners.' An early motorised visitor was none other than the Duke of Bedford, who made a chauffeur-driven excursion from Endsleigh, his Devonshire estate; for an acclimatisation aficionado of his calibre, the lure of Moresk must have been irresistible. After thriving at the site through much of the twentieth century, the Treseders moved away in the mid-1970s, the soil by then exhausted by many decades of cultivation and overrun with weeds. Moresk was later sold for housing. While most of the ornamental plants are gone – save the odd escapee laurel, rhododendron or bamboo – a few of their invertebrate stowaways have lingered.

'How many different flatworm species could we see today?' I asked Dave, as we made our way down into the sheltered valley, each of us haloed by midges. Along with a nine-pronged stainless-steel hand rake, my companion carried a pink bucket containing a blue foam kneeling pad, specimen tubes and head-torch. A camera bag was slung over one shoulder.

'There's about 12 or 14 in Britain,' he said. 'I found a rare one the other day. Probably arrived in the leaf axils of *Phormium*, the New Zealand flax.'

We had been walking for less than a minute when Dave dropped to the ground and started lifting rocks. Moments later he gestured at a shiny black worm, whose back, on closer examination, was revealed to sport a pair of narrow pale grey lines. The thing was barely a centimetre in length.

A couple walking their bulldog paused to see what all the fuss was about, then moved off, doubtless none the wiser.

'*Kontikia ventrolineata*,' announced Dave. 'A semi-adult, that one. It will at least double in length.'

This worm hailed from Australia and, although first recorded at a Liverpool hothouse in the 1970s, is now widespread in southwest England; its presence at Moresk hinted at an earlier introduction courtesy of the Treseders. Little was known of *Kontikia*'s biology save an apparent predilection for small slugs, snails and caterpillars. Nevertheless, the British government was sufficiently concerned to include it in an inventory of four non-native flatworms whose release in the wild was prohibited.

I asked whether we might see the New Zealand flatworm, another species on the banned list, and easily the most notorious.

'There's probably as much chance as winning the lottery,' said Dave, adding, 'There's a few in Devon.'

Compared to *Kontikia*, the New Zealand flatworm, scientific name *Arthurdendyus triangulatus*, is a monster, with adult specimens reaching up to 20 centimetres, and its method of reproduction – known as spontaneous caesarean birth – is pure science fiction. Every ten days or so, the worm's back rips open to release a small, shiny, currant-like egg capsule, containing six or seven young, and then immediately heals over again. Native to New Zealand's South Island, the species was first officially recorded in Belfast in 1963, possibly concealed in a shipment of daffodils, roses or rhododendrons, and two years later was confirmed in the Edinburgh Botanic Gardens. (Workers had in fact observed, but not reported the flatworm, in Edinburgh during the 1950s, and in all likelihood the thing had been arriving for years before anyone noticed.) Since then *Arthurdendyus* has spread, via plant containers, across Scotland, Northern Ireland, parts of northern England, and even on to the Faroe Islands, whose cool, damp conditions resemble those of its homeland. Once restricted to gardens, the slimy intruder is also now turning up on agricultural land – perhaps transferred between farms in hay and silage bales. That's a problem, because it eats earthworms.

The ruthlessness of its predatory behaviour is matched only by its messiness: *Arthurdendyus* ambushes earthworms on the surface by night, or sometimes pursues them down old root channels (*Arthurdendyus* is itself unable to burrow), before wrapping its ribbon-like body around the prey and exuding an anaesthetic enzyme to digest it from the outside in. It hoovers up the resultant goo, leaving behind only a sliver of soil: the earthworm's last meal. And, over time *Arthurdendyus* can annihilate native earthworm populations. A long-term experiment in flatworm-infested parts of Ireland found a 75 per cent reduction in the biomass of what are called anecic earthworms: in other words, the large, vertical-burrowing species, including the familiar lob worm *Lumbricus terrestris*, whose effects on soil structure earn them the accolade of 'ecosystem engineer'. Without the aerating activities of the earthworms, soils become waterlogged and lose fertility, lowering crop yields. In Northern Ireland alone, annual damage wrought by the flatworm was recently put at £34 million. Other animals dependent on the earthworms, such as birds and badgers, may be affected; in parts of western Scotland, the New Zealand flatworm is blamed for an apparent decline in mole populations.

Flatworms weren't the only invertebrates likely to have hitched a ride on Treseder shipments. As we disturbed the leaf litter at Moresk, shiny little black jobs would start bouncing all over the place.

'Landhoppers,' said Dave, as I tried to catch one. 'They're quick. There it is.'

'Got him!' I said, pouncing on a rock, then lifting my hand to reveal the momentarily stunned creature. It was crescent-shaped, resembling a giant flea.

'Yeah. That's an amphipod from Australia.'

Amphipods were a type of small crustacean, usually marine. 'How do you know it's not a native one?' I asked.

'In Britain, the native amphipods we have are aquatic, or live close to water, like sandhoppers on the beach, but we're more than a mile from the sea.'

Also known as lawn shrimps, landhoppers were another southern hemisphere species, and were first noticed during the 1920s jumping about in moist humus and dead leaves in the gardens of Tresco Abbey on the Scilly Isles. No one knows for sure how they got there, but circumstantial evidence again implicates the Treseders. For a start, the Tresco landhopper was native to the nurserymen's collecting grounds of New South Wales and Queensland. What's more, in the years leading up to its discovery, many of the ornamental plants grown in the abbey gardens had been imported via Moresk. And most damning of all is that landhoppers just love tree ferns, with more than 3,000 per square metre recorded among *Dicksonia antarctica* leaf litter at the National Trust's Trelissick Garden in South Cornwall.

Several other species of landhopper have turned up in heated tropical glasshouses, but so far only the Tresco variety has proved tough enough to survive outdoors in the UK. Today, they're found in isolated populations, typically parks and ornamental gardens, across southern Britain and Ireland, including central London: the Natural History Museum's wildlife garden has them, so has Battersea Park and Kew. Landhoppers are capable of spreading under their own steam, covering up to 40 metres in a single night, but their patchy distribution suggests they will hitch a ride where they can.

One thing Dave and I didn't see, perhaps because the weather wasn't right, were stick-insects. Several kinds, originating from

New Zealand, passed through Moresk during the early twentieth century and have now established across parts of Devon, Cornwall and the Isles of Scilly where they feed on various sorts of plants. These include the prickly stick-insect, named for the numerous black spines dotting its thorax and abdomen, and its less spiky relative, the unarmed stick-insect. There is also a smaller species, the smooth stick-insect. A fourth variety, the Indian stick-insect, is found in southwest England, but, as its name suggests, has a different story, having originated in Asia. Otherwise known as the laboratory stick-insect, it's the variety most often kept as a pet or as an educational prop in schools. Unlike its hardier New Zealand cousins, the Indian stick-insect tends to be killed off by British winters. (A couple of European stick-insect species have also naturalised in Britain, with small populations in the Scilly Isles, Hayling Island and Slough.)

All stick-insect species found in Britain can reproduce parthenogenetically, the females laying several hundred fertile eggs per year without need of males. The eggs each hatch into a clone of their mother. In fact, until recently not a single male prickly or unarmed stick-insect had ever been found, even back in New Zealand. Then, in 2016, Dave Fenwick's partner discovered an unarmed stick-insect, measuring 75 millimetres in length, on the side of her car. On closer inspection of the abdomen, Dave thought it might be a male, so too did Malcolm Lee, Cornwall's resident stick-insect expert. DNA testing of the specimen by initially sceptical Kiwi scientists confirmed the suspicion, making headlines across the world. It seems that every so often a mutation occurs and a male pops out; whether or not he gets to mate is yet unknown.

The asexual mode of reproduction explains how a handful of stick-insect pioneers, and even a single egg, can form new colonies, but these masters of camouflage aren't likely to become a problem any time soon. Their flightlessness curtails their ability to spread, and often multiple generations will use the same plant for many years. And, as Malcolm Lee later told me, up to 99.5 per cent of stick-insect hatchlings fail to make it to breeding age. 'If the mortality rate dropped just a tiny bit to 99 per cent, we would have a doubling of the population each year. In a decade, the population would increase a thousandfold!' said Malcolm. 'That we are not up to our necks in stick-insects suggests that the local populations are reasonably stable.'

Are Moresk's non-native invertebrate hitch-hikers worth worrying about? Perhaps. Unless *Arthurdendyus* makes an appearance here, the current crop of flatworms seems fairly innocuous, but what might be their long-term impact on fellow denizens in the leaf litter? Landhoppers are already known to be prodigious feeders – a study in an Irish pine forest calculated that landhoppers consumed a quarter of all leaf litter – so there's a real possibility they're outcompeting and supplanting native detritivores. Stick-insects, being generalist feeders, don't present much of a threat to rare native plants, and most quickly die off, but if average ambient temperatures were to rise, even a bit, all bets are off. These may be concerns for the future, but right now a number of unwanted hitch-hikers are already making their presence felt in Britain.

The figures change all the time, but the most recent estimates suggest that some 400 types of non-native terrestrial invertebrates, mostly insects, are now established in England, Scotland and Wales, of which about a quarter have a negative impact, such as attacking agriculture, forestry or horticulture

crops. Within the agricultural sector alone, non-native invertebrate pests have been blamed for £129 million in yield loss
every year, with more than £26 million spent annually on
trying to control them. And they keep coming: between 2000
and 2010, insects and other invertebrates represented more
than 60 per cent of non-natives reaching Britain, reflecting an
ever-increasing trade in the plants, plant products – from grain
and fruit, to timber and cut flowers – and other goods which
convey them. As we saw with the Asian hornet, the perceived
threat that invasive invertebrates pose to Britain's economy
and environment, not to mention public health, has prompted
the government to implement extraordinary surveillance and
control measures. Since the Tetbury incident, there have been
(at the time of writing) five further nests discovered and
destroyed in southwest England, and separate hornet sightings
in locations from Kent and North Somerset to Scotland.

The oak processionary moth – OPM, for short – is a similarly troublesome recent arrival. The insect, whose adult form
is brown and somewhat unremarkable, originated in central
and eastern Europe, and first established in west London in
2005. It has since spread across the capital and into neighbouring counties. The OPM's larval stage is the problem:
hundreds of caterpillars, which hatch in late spring and early
summer, congregate in the long foraging lines, or processions,
for which the species is named, stripping bare the tops of oak
trees, as well as any beech, birch, hazel, hornbeam or sweet
chestnut in the vicinity. Worse still, each caterpillar is carpeted
in thousands of tiny defensive hairs containing an irritant
protein, which causes itchy rashes on contact with human
skin. The hairs, known as setae, also fall off and get inhaled,
triggering respiratory problems.

Although the odd male OPM is known to have fluttered

across the English Channel since the early 1980s, the present outbreak has been traced to a consignment of infested oak saplings from the Netherlands. As a result, all such shipments now need a special plant passport to enter the UK. The Forestry Commission has meanwhile set up a buffer zone to monitor OPM spread, using pheromone-laced moth traps, and has asked the public to look out both for the caterpillars and the distinctive silken nests in which they aggregate prior to pupation. But control is not straightforward. The most effective current treatment involves spraying infested trees with a pesticide containing a toxic bacterium, *Bacillus thuring-iensis*, but this risks harming native invertebrates, including rare butterflies and moths. There is also a fear of undesirable knock-on effects for the wider ecosystem: when OPM-infested woods in Berkshire were sprayed from a helicopter in 2013, breeding activity among insectivorous blue and great tits appeared to reduce, although a follow-up study found 'no clear wildlife impact'. In sensitive habitats, such as Richmond Park, nests are often simply removed from the trees manually. Ultimately, climate is likely to be the best check on the OPM advance, with a harsh British winter usually enough to knock out colonies.

Another moth introduction illustrates how even non-natives can't escape their enemies for ever. Its name, the horse-chestnut leaf miner, tells you all you need to know, although a 2013 article in the *Independent* newspaper spells it out, describing how the insect is 'wreaking havoc with our conker trees'. Originating in the Balkans – like horse-chestnuts them-selves, which were introduced to Britain in the early nineteenth century – a particularly aggressive strain of the leaf miner moth began spreading north and west from Macedonia, in the former Yugoslavia, during the 1980s. The rate of dispersal, up to 60 kilometres annually, suggested that the leaf miner adults, pupae or larvae were hitching a ride on cars, trucks and trains. A European Environment Agency report notes that: 'It is impossible to park your car under an infested chestnut tree and not translocate dozens or even hundreds of this creature when continuing your journey.' The pest was recorded in Austria in 1989, and – presumably benefiting from Germany's autobahn system – Bavaria three years later. By 2002, it had reached Britain.

Like the OPM, the adult horse-chestnut leaf miner is a small, innocuous-looking thing, although its pattern of black and white chevrons on a red-brown background is more attractive. Again, it's the caterpillars that do the damage, tunnelling within leaves and causing them to scar and drop. Leaf miners don't directly kill horse-chestnuts, but with up to two million larvae infesting a single tree, they will weaken their host, leaving it vulnerable to frost, or diseases such as bleeding canker. This in turn prompts local authorities and other land-owners – fearful of the legal consequences of falling branches – to take the drastic step of removing affected trees altogether. When Wandsworth Council in London felled an historic avenue of 51 horse-chestnuts in 2017, citing 'a serious threat to the public', the *Daily Mail* newspaper dubbed it the 'Tooting chainsaw massacre'. As with the OPM, little can be done to arrest the spread of the leaf miner – which also attacks Norway maple and sycamore. Pesticides or pheromone traps usually prove too expensive, ineffective, or 'ecologically questionable', as one expert puts it. But perhaps the many epitaphs now being written for the much-loved conker tree are premature: early trials of a new biological control agent in the shape of an insect-killing fungus called *Beauveria bassiana* are showing promising results.

It's not only Britain's deciduous trees that are at risk from invasive insect hitch-hikers. As its name suggests, the western conifer-seed bug, a native of North America, has a preference for pine and Douglas fir trees, whose reflected radiation it homes in on like a heat-seeking missile. The three-centimetre-long insects, with a characteristic white zig-zag marking on their back, are thought to have reached Europe several times. The first introduction occurred in 1999, when they crawled out of wooden crates unloaded at a northern Italian airport, the local pine nut industry soon beginning to feel the effects. Since then these adaptable, strong fliers – they buzz like a bee on the wing – have been spreading steadily north and are now present across the south of England. Although primarily a threat to forestry, the bugs will venture into houses to escape cold weather and, if roughly handled, emit a foul-smelling odour. They're even known to bite. Meanwhile Britain's gardens are also under attack from a host of exotic invertebrate pests introduced via the horticulture trade, with the Asian box tree caterpillar and South American fuchsia gall mite among recent and fast-spreading arrivals.

Most non-native creepy-crawlies sneak in unbidden, but a few get released on purpose. For instance, in 1975 several species of earthworm from Mount Kosciuszko, Australia's highest mountain, were introduced to the Scottish Highlands. It was hoped they might do a better job than the native earthworms at decomposing the peat, but the Aussie worms had little impact. At least one type is said to linger on. More frivolous reasons can underlie invertebrate introductions, as evidenced in the Victorian and Edwardian craze for 'butterfly-planting', which saw the insects deliberately released in new sites around Britain. Those responsible often acted out of simple curiosity or perhaps felt they were improving the countryside (another manifestation of acclimatisation). A few butterfly-planters though, may have had less honourable motives, hoping that the confirmation of a desirable variety at a new location would raise its price or prompt a rush of collectors to the area, boosting local tourism.

One early example dates to 1855, when Joseph Merrin, editor of the *Gloucester Journal*, was shown four specimens of large copper, a spectacular species that had recently died out in the fens of East Anglia. Merrin was informed that the butterflies had been netted 'sometime previously on the slopes of Doward Hill, bordering the River Wye, not far from Monmouth'. Convinced of the record's validity, the newspaperman went looking in vain for the insect. In a separate case from 1912, the European map butterfly – a beautiful variety known only from the continent – was released in Gloucestershire's Forest of Dean. Unlike the large copper, the European map, whose caterpillars feed on stinging nettles, quickly established itself in the forest, as well as in nearby Symonds Yat, much to the delight of local entomologists. But Albert Brydges Farn, a septuagenarian former public health administrator and proud owner of Britain's finest private collection of native butterflies, was having none of it. Ahead of his time when it came to attitudes to 'alien' flora and fauna, Farn was horrified to learn that a foreign species was at liberty in the British countryside and resolved to single-handedly exterminate the colony. Despite failing health, he spent three years secretly clambering around the rugged hills until the job was done.

Expert bug-hunters believe that butterfly-planting isn't just a quaint historical practice, but something that persists to this day. Entomologist Richard Comont sees something fishy about a recurring population of marbled fritillary at Fineshade Wood in Bedfordshire: 'This is a continental European species,

never been seen properly in Britain and not a migrant species.'
Then there's the Glanville fritillary. Small numbers naturally
occur on mudslopes on the Isle of Wight and the Channel
Islands, where its larvae feed on plantains, but Richard suspects
human agency when it comes to another population at
Weston-super-Mare in North Somerset: 'Suddenly one year
there's loads of Glanville fritillary, then they gradually decline.
It's because someone drives there with a carboot-full of butterflies and lets them fly out. It's been happening year after year
for at least eight years.'

At least most invertebrate hitch-hikers can be detected if you
look hard enough. But we also face a barrage of bacteria,
viruses, fungi and other microbes which are all but invisible
to the naked eye. Among the deadliest of these belong to a
poorly known group of microscopic organisms called oomycetes (pronounced *oh-oh-my-seet-ees*). Their common name,
water moulds, suggests them to be a kind of aquatic fungus.
In fact, oomycetes are more closely related to algae, and not
all inhabit moist conditions. Most of the several hundred
species of water mould are harmless and play a useful role
in helping to break down organic matter, but a few give the
group a bad reputation. These include the water moulds
responsible for sudden oak death and crayfish plague, along
with those causing diseases in farmed salmon. Downy mildew
in cucumbers and watermelons is also the work of an oomy-
cete. But none has had more profound consequences than
Phytophthora infestans, commonly known as the potato late
blight.

The spores of this most infamous of water moulds develop
on the host plant's leaves before being washed down into the
soil to infect the growing potato tubers. When harvest time

comes around, all that remains of the spuds is a sticky black mess. The strain responsible for the Irish Potato Famine originated in Mexico and is thought to have been imported to Europe in the 1840s on diseased potatoes via North America. Although southern Britain and the Low Countries were also affected, Ireland came off worst thanks to a damp maritime climate and reliance on a single strain of potato, the 'Lumper', with no resistance to the blight. As much as half of the Irish crop was lost in 1845, the first year of the outbreak, with further episodes in 1846 and 1848. The situation was exacerbated by a dysfunctional political and economic system, which saw Ireland exporting an abundance of agricultural produce to the rest of the United Kingdom (of which it was then part) while its own people starved. As a result, a million Irish, equivalent to an eighth of the population, perished from hunger and associated epidemics, with further millions emigrating. Chemical treatments and blight-resistant potato varieties are now available, so another catastrophe on this scale seems unlikely. *Phytophthora infestans* nevertheless continues to be a significant problem, causing an estimated $6 billion of crop damage every year, and new versions are evolving all the time, including 'Blue 13', an aggressive strain identified in 2005 and now Britain's predominant blight.

Sometimes hitch-hiking invertebrates and plant pathogens can combine to deadly effect. This is best evidenced by Dutch elm disease, a fungal infection, whose spores are spread by several species of bark beetle. Thought native to Asia (where the elms seem resistant), the fungus, scientific name *Ophiostoma ulmi*, first came to Europe from the Dutch East Indies, present-day Indonesia, in around 1910. It rapidly spread across the continent, causing some countries to lose up to 40 per cent of their elm trees. The disease later crossed the Atlantic, probably in a consignment of contaminated elm veneer logs, going on to devastate American trees.

Then, during the 1940s, an entirely new, more aggressive, form of the fungus, *Ophiostoma novo-ulmi* – also spread by beetles – emerged in the Great Lakes region of North America to deliver a sort of *coup de grâce*; within a couple of decades, a once-common ornamental tree had all but gone from the eastern seaboard of the United States and Canada. In the 1970s, this more virulent strain made its way back over to the UK in diseased logs used as dunnage in ships. Lowland central and southern parts of Britain were hardest hit by this second onslaught, and within a decade the disease had put paid to two-thirds of the country's 30 million elms. (Curiously, a different subspecies of *Ophiostoma novo-ulmi* also appeared in the Moldova–Ukraine area of eastern Europe during the 1940s, spreading east and west, and eventually overlapping, and hybridising with, the American strain.)

Today, just a handful of mature elm trees remain in the UK, and these are protected behind *cordons sanitaires* in Brighton and Edinburgh, where intensive management efforts are deployed, such as the rapid felling and burning of diseased elms, and the digging of trenches between neighbouring trees to prevent fungal transmission via root systems. Dutch elm disease doesn't always kill the entire tree: a few roots may survive, so that every spring, hedgerows across Britain host a macabre cycle of birth and premature death, as vigorous thickets of elm saplings briefly sprout only to be cut down by new waves of infection.

Memories of this devastating outbreak had barely faded when another of Britain's most important native trees faced an onslaught from its own new fungal hitch-hiker. First documented in Poland in 1992, ash dieback – formerly known as *Chalara* – causes infected ash trees to drop their leaves, as lesions form on the bark. The tree succumbs either directly from the infection or from secondary attack by other pests and pathogens, notably honey fungus. Saplings under ten years of age are the most vulnerable, but ash dieback is quite capable of killing a mature tree. The disease's presence in Britain was confirmed in February 2012 at a nursery in Buckinghamshire, having arrived there in a consignment of 2,000 plants from Holland. A ban on all new imports of ash was triggered, although by then other nurseries had probably also received contaminated shipments. By December 2014, almost a thousand outbreaks had been recorded, predominantly across the east of the country, as fungal spores were released from infected fallen leaves and carried on the wind.

Today, ash dieback is present in half of the UK with government scientists predicting that Britain could eventually lose almost all of its 90 million ash trees. Given that ash compose 15 per cent of broadleaved woodlands, the change to the landscape would be profound. Recent research indicating that British trees might be more tolerant to ash dieback fungus than continental counterparts has offered a glimmer of hope. Unfortunately, the same genetic factors conferring resistance also render a tree more susceptible to insect predators. That's a big problem, given that a new, and even more potent, enemy of our embattled ash is now on the horizon.

Beautiful it may be, but the emerald ash borer has been a calamity for North American ash trees. This vivid green beetle, whose larvae suck the sugar-rich phloem sap from their host, is native to China, Taiwan, Korea, Mongolia and the Russian far east. Within a few years of its 2002 arrival in Michigan, six million of the state's trees were dead or dying. Back in

Asia, where ashes have had longer to evolve resistance to the beetle, only stressed or dying trees are affected, but in its new realm this natural born driller, capable of attacking healthy trees, is having a field day.

By 2014 emerald ash borer, whose adult stage can fly long distances, had killed 200 million trees across 22 states, making it the continent's costliest forest pest. Hopes are now resting on biological control: trial releases have taken place of several species of tiny wasp and a fungal pathogen from China known to kill emerald ash borer larvae, although there's a concern they might impact native beetles too. In the meantime, Britain's plant health scientists are on high alert, fearing that the beetle might one day hitch-hike to Britain on consignments of firewood from Ukraine or the Baltic. As one government scientist told me, 'If you think ash dieback is bad, this is much worse. The emerald ash borer is no Mickey Mouse.'

Throughout human history, hordes of invertebrate pests, pathogens and other undesirable stowaways have dispersed themselves by hitching a ride on plants and their products. But all the while, a special group of plants has turned the tables by exploiting animals to criss-cross the planet. Botanists call them wool aliens and, to see some for myself, I headed to Newton Abbot in South Devon.

Roger Smith of the Botanical Society of Britain and Ireland (BSBI) met me at the railway station in his red Ford Fiesta. The mid-August traffic was appalling, a combination of the summer holidays and it being raceday at the town's racecourse. A quarter of an hour later, having covered barely a kilometre, we parked up near a terrace of Victorian cottages. Ahead was a cluster of large, disused red-brick buildings – the Bradley Mills.

'As it gets more developed there's less and less stuff to see,' said Roger, who was BSBI's County Recorder for South Devon, but after a few paces he pointed down at something. 'There's one.'

Roger was referring not to a pretty blue flower near the front door of one of the cottages, but to a scrubby little thing next to it.

'*Cotula australis*. Can't remember where it's from: New Zealand or Australia, I think,' he said.

Opposite the cottages was a narrow channel of water, clogged with algae and litter; all that remained of a leat diverted from the town's River Lemon to power an earlier incarnation of the mill. We approached the abandoned factory, its doorways and ground-floor window spaces permanently sealed with red bricks or grey breeze blocks. A plaque read, 'J. V. and S. rebuilt 1883', denoting a restoration by the then owners, John Vicary and Sons, following an inferno. This was in fact the fourth time the mill had been destroyed by fire in its long history.

A clump of grass caught Roger's eye.

'Water bent, *Polypogon*. That was on Mary's list. Almost certainly a wool alien.'

Wool aliens. The name suggests a sort of mutant sock-puppet from Mars, but Roger was referring to a class of non-native plants whose presence in Britain is almost certainly due to the wool trade. These globetrotting botanical wonders had evolved spines, bristles, barbs, hooks, burrs, gluey secretions called mucilage, and other modifications of the fruits and seeds, which allowed them to catch a ride on shaggy livestock or their fleeces. Wool aliens thus travelled the world.

From the eighteenth century, as Britain's textile industry grew, demand for raw wool far outstripped home-grown production, forcing mills to import vast quantities from Australia, New Zealand, southern Africa and the Americas. The fleeces would be scoured clean at high temperatures, with strong alkalis and acids, and the greasy seed-laden residues, known as grey shoddy, dumped in piles outside the mill. Despite this harsh treatment, seeds from as far afield as Chile, Cape Town and New South Wales began germinating on shoddy heaps the length of Britain. Rich in organic material, the shoddy was also applied as fertiliser to agricultural land and orchards, further distributing the wool aliens.

Botanists started to take an interest in wool aliens after a pioneering study conducted by Ida Hayward of the Linnean Society at Galashiels in the Scottish Borders. Long associated with the wool trade, by the late nineteenth century Galashiels was booming, as its mills churned out yarn, cloth, stockings, shawls, blankets and myriad tartans, while disgorging untreated wool scourings into the nearby River Tweed and Gala Water. Between 1908 and 1917, Hayward, whose uncles worked in the local industry, identified 348 non-native plant species along these watercourses. Almost half originated in Europe and the Middle East, but a significant proportion were from the southern hemisphere. Similar assemblages were noticed in other hotspots of the wool industry, such as Yorkshire: in 1952 a single botanist collected more than 40 species in Bradford. By the 1960s, more than 500 kinds of wool alien had been counted in Britain.

It wasn't all one-way traffic: some of the most tenacious wool aliens in Australia, such as false or wall barley *Hordeum murinum*, probably originated in Europe. At school, we would derive amusement in tossing its neatly detachable and pleasingly aerodynamic seed heads at each other; that these missiles were known as 'flea darts' added to the fun. But for Aussie farmers, false barley is no joke: its sharp points can injure, even kill, livestock. Similarly, horehound *Marrubium vulgare*, a member of the mint family first introduced from Britain to Australia as folk medicine in the nineteenth century, now infests 26 million hectares, including some of the Outback's remotest sheep stations, thanks to the propensity of its seed-bearing burrs to cling to wool.

Britain's wool aliens, like the mills, tended to be concentrated in the north – the moister conditions better suited textile processing – but Newton Abbot's Bradley Mills once boasted its own modest assortment. In November 1957 a Scottish botanist, Mary McCallum Webster, documented every last one she could find in the vicinity. Her list, published two years later in the *Report and Transactions of the Devonshire Association for the Advancement of Science, Literature and Art*, ran to 57 species. She recorded eight more at nearby Buckfast, the site of another mill. A few years later, another botanist visiting Bradley shoddy dump found a further 14 non-natives.

McCallum Webster's original tally included aromatic peppercress *Lepidium hyssopifolium*, an Australian member of the mustard family with sticky seeds; and several species of *Erodium*, a type of geranium known as storks-bills for the twisty beak-like fruits which help them cling on to sheep's wool. Among the more sinister of McCallum Webster's finds, the yellow unicorn plant *Ibicella lutea*, wasn't strictly a wool alien. The seed of this South American native, which she described as 'a fearsome looking object with a hard warty body about three inches long and two curved horns about eight inches in length', has evolved to lodge, painfully, into a sheep's foot. The unfortunate animal not only provides transportation but nutrition too, for when the conscripted chauffeur dies, the seeds, nicknamed 'devil claws', germinate in its rotting corpse.

Roger and I found no devil's claws that day, and indeed little else. We weren't surprised. Belying their tenacity, many wool aliens tend to be transient casuals: they germinate then die out, so require constant replenishment. As Britain's wool industry declined in the middle of the twentieth century, the textiles market was flooded by cheaper Asian imports, so the exotic flora quickly faded. An Australian botanist visiting Galashiels in the mid-1960s found no trace of Hayward's wool aliens; in fact, they'd long before been exterminated by a combination of Scotland's harsh winters and new drainage systems to intercept and treat the mills' wastewater. Given that Bradley Mills had closed its doors for the last time in 1972, it was a testament to Devon's friendlier climate that Roger and I had found any at all.

But not every wool alien has done the decent thing and died out. In 2010, almost 30 new species appeared when soil was ploughed at a farm in Flitwick, Bedfordshire. It transpires that the shoddy had been used there back in the 1980s, and the aliens had been lying dormant ever since, just waiting for their moment in the sun. Pirri-pirri burr, meanwhile, a member of the rose family native to Australia and New Zealand, is positively rampant. First recorded growing wild in Britain in 1901 – and also featuring on Ida Hayward's list – dense mats of the plant, recognisable by its spiny red seed heads, are found along the Tweed. It also infests the sand dunes of Lindisfarne island where it threatens a rare endemic orchid, while in the RSPB's Minsmere nature reserve in Suffolk, pirri-pirri burr causes a major nuisance and worse when it gets tangled in birds' feathers.

It's a bit of a stretch, but we might also apply the term

'wool alien' to an altogether different assortment of exotic species; a group of animals whose intelligence, adaptability and attractiveness to humans has made them among the UK's most successful colonists – the mammals.

8

Fur Farm

'From the desperate city you go into the desperate country, and have to console yourself with the bravery of minks and muskrats.'

Walden; or, Life in the Woods,
Henry David Thoreau, 1854

Dudley, West Midlands. August 2002

It had been a tiring few hours. We had reached Dudley at around 3 pm, having driven up the motorway from our base in Bristol. My companions, a senior producer at a major television broadcaster – let's call him 'Mike' – and our cameraman 'Paul' had dropped me with a quarter of a tonne of lighting equipment at a sports ground, before proceeding to the town's zoo. My task, as assistant producer, was to lay heavy-duty cable from a cricket pavilion (the power source) several hundred metres up the steep side of a nearby wooded hill. I was also charged with hauling lamps, tripods, ballast units and other weighty paraphernalia to the summit.

It was sweaty work, not helped by waist-high stinging nettles and clouds of biting flies. But I made good progress: by dusk, the lights were positioned in a forest clearing, and Mike and Paul were back from filming at the zoo. The goal of today's activities was to get 'pick-ups'; in other words, a few important filler shots to enable us to complete a natural history documentary sequence, most of which had been filmed a few weeks back on location in Chicago. The sequence was about raccoons, revealing the ways these intelligent and inquisitive nocturnal omnivores exploited North American cities. One focus had been their dexterous and sensitive front paws, which could as readily extricate food from a suburban dumpster as snatch crayfish from a mountain stream. Just a handful of action shots were left to do, depicting an animal skulking through a forest at night – and a Dudley forest would serve the purpose. All we needed now was our performer, and then it was a wrap.

Right on cue 'Dave', the animal wrangler, and assistant 'George', appeared from behind a tree. George was carrying a cage, its occupant a feisty-looking pet raccoon called Albert (again, not his real name). At the time, raccoons were classed as 'dangerous wild animals' under British law and anyone wishing to keep them required a licence from the local authority. I wasn't *au fait* with the terms of Dave's licence, but I imagined an important stipulation was that on no account should a raccoon be released into the wild. Dave though seemed relaxed about this, no doubt confident he had instilled the necessary discipline into his animal, which after all was surely the veteran of earlier nature documentaries and perhaps some television commercials to boot.

Paul was ready to roll. I threw some switches and our Arri lamps blasted the clearing with ten thousand watts of cold white illumination (simulating moonbeams takes a fair punch of light). From a distance, it must have looked like a scene from *Close Encounters of the Third Kind*; not so far from the truth given that an alien would soon be on the loose.

At Mike's signal, George slipped the bolt on Albert's cage. Seconds later, in a bushy blur of black and white, our raccoon was gone, lost into the shadows.

'Oh, *no!*' exclaimed Dave in his West Midland cadences. 'If we don't catch him, I'm in the *sheet!*'

Mike, Paul and I suppressed smirks as the animal wranglers scoured the hillside for the fugitive. Half an hour later, when it became clear that Albert wouldn't be returning to the set, we packed up and headed home. Not a single frame of 'raccoon creeps through forest' made it to the can that warm summer night, and whether Albert ever rematerialised remains a mystery to this day. (Although, as recently as September 2011, a tame raccoon was said to be haunting a Dudley garden.)

Albert wouldn't be the first of his kind to gain the run of the British countryside. Raccoons have been spotted on and off since the late 1970s, but until now they haven't got going in the wild. It's a different story on the other side of the Channel, where the North American carnivore has established colonies from Spain and France right across to Belarus and Ukraine. Germany is a particular stronghold, boasting upwards of half a million raccoons. Many of these continental populations were started by escapees from fur farms, zoos and private collections, although in Germany, Poland and Russia, raccoons were deliberately introduced for reasons of 'faunal improvement'; also, during World War II, a number of raccoons, kept as regimental mascots in the US Army, are rumoured to have escaped into the European countryside.

In the UK, rules for keeping raccoons were relaxed in 2007: no longer would keepers need a licence. That seems to have been a mistake. With their popularity as pets rising ever since, raccoons have been listed as among the top ten most ecologically harmful exotic species likely to naturalise. That they haven't already done so is down to chance alone. Raccoons prey on native birds and amphibians, raid agricultural crops and host multiple parasites and pathogens, from roundworm to canine distemper virus, so could cause mayhem if they ever established themselves in the wild. In 2016, the European Commission belatedly included them on a list of 37 species now illegal to import, keep, breed or release into the environment. A government official told me that most raccoons seen in the wild tend to be escaped pets, adding that the northeast got a lot of sightings. 'Perhaps they're Newcastle football fans,' he jested, a reference to their black-and-white markings.

If raccoons ever do get their act together and colonise Britain, they'll join a growing list of non-native mammals already roaming this country: getting on for a third of our terrestrial species are classed as introduced, naturalised or feral. A scientific paper, published in 2012, ranked the UK as Europe's third most mammal-invaded region, whose tally of 18 established species was beaten only by two other islands, Sardinia and Corsica (with 21 and 19, respectively). In fact, Reeves' muntjac, Chinese water deer and Père David's deer were omitted from the score, so perhaps Britain has a fair shout at the top spot. However you look at it, right now we're living through a golden era of mammals, with our furry fauna richer than at any time since the Neolithic period, albeit with a few of the larger carnivores now absent.

Why do introduced mammals do so well in Britain? Well, for a start they all bring along their own supply of nutritious food – milk – for their youngsters. The most successful mammal invaders, including rats, mice and cats, also tend to be intelligent and opportunistic, enabling them to adapt to new environments. Climate doesn't put them off either: an ability to internally regulate their body temperature and, if necessary, hibernate sees mammals through the harshest of conditions. And the same fur that keeps them toasty goes a long way to explaining why so many have been brought here on purpose.

Throughout the history of human settlement on these chilly islands, we've exploited native mammals as much for their fur as for their flesh, in the process wiping out, among others, brown bears, wolves and beavers. (The latter were also targeted for castoreum, a secretion of the scent glands highly prized for its medicinal properties and as a base for perfumes.) From the beginning, fur offered status as well as warmth, with the finest pelts reserved for the wealthiest and most powerful – the value of certain skins only climbing as the animals which bore them headed for extinction. In the tenth century, as beaver, marten and ermine numbers dwindled, their coats were reserved for the king alone, the plebs having to make do with cheaper alternatives such as goat and sheepskin. Little wonder that entrepreneurs were always on the lookout for new fur-bearers that could be domesticated, rabbits being a prime early example. By the twentieth century, fashion not function powered the fur trade, both in the UK and abroad, with a select handful of species accounting for most of the production. Those same animals are today ranked as some of the world's worst invasive mammals, from the North American beaver in Tierra del Fuego to the Australian possum in New Zealand. In Britain, the late 1920s ushered in a trio of fur-bearing New World invaders. Their careers would follow very different trajectories.

The muskrat, a stocky cat-sized rodent with a long, rudder-like tail, offers a textbook example of biological invasion. Even before the international fur trade found its feet, this amphibious mammal was widespread, with a natural range stretching from Alaska to northern Mexico, and all points between. Today, it's truly cosmopolitan, with non-native populations extending from France to China and Japan, as well as across parts of South America. In 1905, a Bohemian aristocrat called Josef Colloredo-Mannsfeld brought back three females and two males from a hunting trip to Alaska, releasing them on his Dobrisch estate near Prague, in the present-day Czech Republic. Four years later, the wild population had multiplied and was spreading so uniformly across Europe that mathematical models could predict its advance. The march of the muskrat has been likened to ripples from a pebble thrown into a pond or, less poetically, to *The Blob*, referencing a 1958 science-fiction movie. Muskrats turned out to be incorrigible pests, raiding fish farms and undermining the banks of rivers, canals and dykes. While fur farm escapes added to the phenomenal success of the muskrat, the rodents continued to be deliberately loosed, notably by another man named Josef. A late convert to acclimatisation – a practice by then discredited – Josef Stalin, the Soviet dictator, ordered the liberation of muskrat, mink, coypu, sika deer and other non-native mammals into the Soviet countryside during the 1920s and 1930s as part of a crazed plan for the 'Transformation of Nature'. He also called for the liquidation of foxes. There was a certain irony to this, since another exotic released into the wild on Stalin's orders was the raccoon-dog, essentially an East Asian version of the fox.

Introduced by breeders to Scotland and Ireland in 1927 and, two years later, to England, the career of muskrat in the British Isles was relatively short-lived. At first, as elsewhere in Europe, the rodent was valued for its waterproof fur, called 'musquash', with government officials encouraging what they viewed as 'a new and useful industry'. Yet, multiple escapes into the wild were soon being reported, with muskrats – nicknamed 'swamp rabbits'– digging in along a 30-kilometre stretch of the River Severn above Shrewsbury. At the urging of mammalogists, who had seen on the continent how the rodents could get out of control, the authorities responded with unusual speed and ruthlessness. From 1932, any fur farm wishing to keep muskrats had to apply for a licence under the Destructive Imported Animals Act; the following year the species was banned altogether.

Meanwhile, an extermination campaign, involving the use of lethal jaw-traps, was launched under the guidance of a certain 'Herr Roith', a Bavarian expert. By the end of 1934, some 4,382 animals had been eliminated in England – more than half from the Severn region alone. In Scotland, where a number of escapes had also occurred, a similar exercise between 1932 and 1935 trapped 958 muskrats. Hundreds more were killed in Ireland. Alas, the traps weren't very selective, with many thousands of water vole, moorhen and other native creatures perishing as collateral damage. The campaign was nonetheless judged to have been a roaring success, with Britain's muskrats all but gone by 1937.

The tale of the coypu – another large, semi-aquatic rodent – is remarkably similar. Once restricted to the waterways of

South America, this walrus-whiskered, orange-toothed beast is now well established around the world, again thanks to the fur trade, which identified it as a promising alternative to beaver. Although the pelt of the coypu, which is also known as the nutria (a term derived from the Spanish for 'otter'), was regarded as inferior to that of beaver, through selective breeding a broad range of colours could be produced. The species was also easier to contain than its northern hemisphere counterpart and, just as important, female coypu reach maturity at just three months (for beavers it takes up to three years) and breed again within a day of giving birth. This made coypu an ideal farmed animal, but also one with enormous invasive potential. As early as 1882, coypu had been brought to France, but the bulk of the western European introductions date to the twentieth century, with animals also farmed, or released for hunting, in the United States, Russia, Kenya, South Africa, Israel and Japan. In North America – where they're known as swamp rats – coypu have since gone on to displace native beavers from wetlands, by eating food faster and even mounting deadly raids on their lodges. The mammal is also wreaking similar havoc in the wetlands of southern France.

In 1932, just three years after the coypu was brought to Britain, the first breakout was reported, in Horsham, West Sussex. By the end of the decade, around half of the country's 49 fur farms then known to keep coypu had lost animals, with escapes reported in Buckinghamshire, Cheshire, Devon, Essex, Gloucestershire, Hampshire, Huntingdonshire and Staffordshire. Not all releases were accidental: the price of coypu fur crashed on the eve of World War II, forcing the closure of farms, with many proprietors disposing of their valueless stock into the wild. Coypu were less tolerant than muskrat of British winters, but the 1950s and 1960s saw an estimated 200,000 ranging across the flatlands of southern and eastern England, chewing their way through fields of sugar beet and Brussels sprouts, and eroding river banks and dykes. In 1961, a long-term coypu-trapping programme was begun, with 97,000 animals killed in the first year. But this did little to slow the species which, by the end of the 1970s, was being blamed for dramatic ecosystem changes. In the Broads of East Anglia, coypu were accused of wiping out certain aquatic plants (although a rise in water pollution and boating activity may also have been responsible). In 1981, all the stops were pulled out with a new £2.5-million extermination campaign launched to finish the job. Eight years later, the coypu became only the second naturalised mammal – after the muskrat – to be eradicated from Britain. (The mammalogist Derek Gow reports, however, that coypu are now in the process of renaturalising in several parts of Ireland, possibly having escaped from pet farms.)

For the American mink, a member of the mustelid family (which also features weasels, stoats and otters), things would turn out differently. Unlike with muskrat and coypu, a Eurasian version of the mink already existed, albeit only distantly related to the New World species. Once widespread across Europe – although apparently absent from Britain for some 500,000 years – by the twentieth century a combination of habitat loss and overhunting for its luxuriously dense winter coat had pushed the Old World mink to the verge of extinction. In response, European fur traders began sourcing its larger and more aggressive cousin from across the Atlantic. As with imported muskrat and coypu, the newcomers were either farmed, or released into the wild for hunting, and soon got a foothold on the continent. Today, the non-native mink is found from Scandinavia to Russia, and south into Spain and Italy, where it is driving out the few remaining European mink.

American mink got off to a relatively sluggish start in Britain. Although farm escapes had occurred within a few years of the 1929 introduction of animals from eastern Canada and Alaska, until 1956 the species wasn't known to breed in the wild. Thus, while timely action was taken against the more prolific muskrat – and, later, coypu – mink farming was given a pass, and the naturalised mink population was allowed to multiply and spread under the radar.

The industry intensified after World War II, peaking in the early 1960s, with 700 establishments across England, Wales and Scotland, producing 160,000 pelts annually. Many of these farms reported losses, while all the time fresh stock flowed in from North America and breeders in Scandinavia. The farmers bred various colour mutations, including light brown, grey and, most prized of all, pure white (produced by mink that were congenitally blind). At first, populations in the vicinity of farms reflected these variations, but natural selection generally saw later generations reverting to the normal chocolate-brown coloration that offered better camouflage in the wild. (One recent exception to the rule was a population of mink, sporting a coat of raven blue-black, that was eradicated in 2007 close to the Duchray river in Perthshire, Scotland.) Not until 1964 did the British government attempt a half-hearted initiative to control wild mink. The effort proved futile and was soon abandoned.

Mink farming eventually took a nosedive during the late 1980s, partly the outcome of a concerted campaign alleging the cruelty of the trade, partly the result of the collapse of the Soviet Union, then a major export market. With businesses folding left, right and centre, unwanted mink were often turned out into the surrounding countryside to save any further expense. A decade on, fewer than a dozen farms still operated. These found themselves subject to the attentions of animal rights activists on a mission to liberate the remaining captives. In one high-profile case from 1998, up to 7,500 mink were loosed from a single farm in Newcastle-under-Lyme, Staffordshire. Despite an intensive round-up operation, many were never to be seen again.

Today's 'mink problem' is often blamed on such episodes, but few of the escapees are thought to have survived long in the wild; and in any case, by then, the species was already entrenched the length of the country, even colonising remote Scottish islands (although mink have now more or less been eradicated from the Outer Hebrides). Moreover, after years of selective breeding, farmed mink by the late 1990s were more docile than initial feistier introductions, so were easier to recapture after releases and unlikely to fend for themselves in the wild. When the last mink farm closed its doors in 2003, the naturalised population stood at an estimated 37,000 animals. Attention was now turning to the impacts of this new, voracious addition to Britain's fauna.

At first, mink were blamed for the disappearance of the otter, a fellow semi-aquatic mustelid, whose populations crashed between the 1950s and 1970s. But the decline in otter was chiefly a result of agricultural pesticides washing into watercourses during this period. Widespread culling for their pelts, and persecution by river bailiffs to protect fish stocks, were other important factors; between 1958 and 1963 alone, more than 1,065 animals were slaughtered by the 11 otter hunts then operating. By the time the practice was banned in 1978, the mammal had all but gone, with mink moving in to fill the vacant niche. Later, when improving water quality boosted

otter numbers, these much larger mustelids aggressively
dislodged the American interlopers. The data is patchy at best,
but in some parts of England, such as Devon and Cornwall,
mink numbers are now falling due to the otter resurgence. The
mink's greatest impacts seem instead to be felt on outlying
Scottish islands, where the species devastates colonies of nesting
gulls, terns, lapwings and redshanks. On the mainland, meanwhile, it's the mink's association with the catastrophic decline
of a much-loved rodent that causes the most consternation.

In the 1950s, the UK's water vole population was put at
around eight million; by 2004, just 220,000 were left. This
appalled the British public, in whose mind the species is affectionately linked to Ratty, a character in Kenneth Grahame's
children's book *The Wind in the Willows*. The mink was not
wholly to blame, since Ratty, who famously loved messing
about in boats, was already suffering from people messing about
with the countryside. It's been estimated that water voles
have lost a third of their natural habitat since 1940 due to
agricultural intensification. Other natural predators, including
foxes, stoats, buzzards, herons and gulls, will take water voles,
but there's no disputing the mink effect: between 1989 and
1998, as mink numbers were peaking, those of the water vole
– a favoured prey item – fell by 90 per cent. You might ask
why the recovering otter population doesn't get the blame.
The answer is simple: unlike mink, an adult otter is too big
to pursue a water vole into its burrow. A cartoon image of
Wile E Coyote chasing Road Runner down a hole and getting
stuck comes to mind.

The River Axe in southwest England is one of my favourites.
Rising near Beaminster in West Dorset and meandering
unfussily for 35 kilometres southwards through Somerset and

Devon before emptying into Lyme Bay, the waterway drains a lowland landscape of understated beauty. The Axe weaves rather than cuts a path to the sea. Its name derives from *isca*, an ancient word for 'abounding in fish'. (This was perhaps not the early Brits at their most imaginative – indeed, North Somerset boasts an Axe of its own.) True enough, the Axe still hosts decent populations of fish including, in testament to the quality of its water, varieties sensitive to pollution, like salmon, sea trout, lamprey and bullhead. These days, the valley's lower reaches are best known for the rarefied birdlife that breeds there, from warblers and water rail to peregrine falcon and barn owl. Great white egrets and osprey are among regular visitors expected to settle in the near future. One species that the Axe until very recently *didn't* boast was the water vole.

'They died out in Devon in 1997. Pollution and other man-made factors would have reduced the population. But it was mink that finished them off,' said 'Richard', a conservationist now working to put that right. (Another bogus name, I'm afraid: Richard's job requires him to kill mink, and some still take exception to that, decades after the last animal activist raid on a fur farm.) In 2009, the local council launched a water vole reintroduction project, with funding from Natural England, releasing an initial 100 captive-bred animals on the Lower Axe. Ever since then, hoping to give the new voles a fighting chance, Richard and his colleagues have been suppressing the North American menace. And that's not easy. The mink is a stealthy predator, tricky to detect, and is firmly entrenched in the region; indeed, some of Britain's earliest wild colonies were recorded in Devon. The current approach centres on the so-called 'mink raft', as I discovered when, one sunny October morning, I met Richard on the nature reserve he manages.

To a soundtrack of trilling skylarks and the occasional dissonant squawk of a wayward pheasant, he guided me down to the muddy edge of a reed-lined stream. Here was a sort of giant plastic shoebox open at both ends – the 'tunnel' – mounted upon a floating rectangle of thick, waterproofed, polystyrene. Richard removed the tunnel to show me a plastic basket containing a block of Oasis floral foam topped with a clay and sand mixture. The basket was set into the raft so that its absorbent foam, constantly in contact with the water, kept the reddish-brown surface damp. Thus, any visitor small and inquisitive enough to explore the tunnel couldn't help but leave its footprints.

'How can you tell when they're mink prints?' I asked.

'Well,' replied Richard, 'they're smaller than otter, bigger than a brown rat. They've got a kind of sharp flowerhead pattern. You get your eye in and spot them pretty well.' Coots, moorhens and shrews were among other animals known to pitter-patter onto the rafts. As Richard explained, four such contraptions were strategically located in various ponds and streams around the site. The rafts were tethered close to the bank, amid the sort of marginal vegetation mink liked to patrol.

When prints were detected, Richard would switch the basket for a spring-loaded cage trap, triggered by a sensitive footplate. The latter was calibrated to the heft of a typical mink, but non-target species could sometimes set it off. ('It's a devil for catching your fingers, as well.') Animal welfare considerations now kicked in: as soon as the primed trap had a customer, Richard needed to know, in order to reduce unnecessary stress to any captured animals – including unwanted bycatches. That used to mean checking it as often as possible, ideally several times in a 24-hour period. To avoid this labour-intensive process,

Richard recently installed 'MinkPolice' equipment. Developed in Denmark, these consisted of small waterproof units attached to each cage, which were activated when an animal was trapped, sending Richard's mobile phone a text alert.

The final stage was delicate, and not for the squeamish. A trapped mink was usually smelt before being seen: when upset, it would produce a musk, which Richard likened to the oily, garlicky, fishy odour of a grass snake when handled. 'A real metallic, tangy scent that lingers in the back of your throat.' Captives were also wont to emit an electrifying high-pitched scream. Richard therefore would do his best to minimise any disturbance, manoeuvring the cage just enough to allow him to rest against it the muzzle of his air-rifle loaded with metal pellets. If all went well, the animal would remain unperturbed, and might even come up and sniff the gun. At other times, it would have to be gently cajoled into position with wooden combs. A single and instant kill shot was the only acceptable outcome. 'You try and do it as quietly and quickly as possible, without rushing, to cause the least distress to the animal,' said Richard. 'You're always going for one clean shot between the eyes or ears.'

Richard usually culled somewhere between 13 and 20 mink every year on the reserve, a statistic that had, until recently, remained stubbornly constant. He is now seeing a decline in mink presence on the river, with just four animals caught so far this year. This has coincided with more frequent sightings and field signs of otter in the vicinity, reflecting the wider pattern across the southwest of England. However, mink numbers on the Axe are unlikely to fall to zero any time soon. The problem is that females, naturally warier than males of the rafts, are seldom caught. So, the mink keeps breeding and fresh recruits are always ready to fill the void.

'Maybe trapping's not the answer?' I asked.

'Yes. But at the moment, it's the best-known form of control,' sighed Richard, who planned to keep going. On the plus side, Ratty is thriving again on the Lower Axe: almost 300 water voles at the last count, with further releases planned. Perhaps Richard was doing just enough.

The fur trade has much to answer for when it comes to problematic non-native mammals, but it cannot be held responsible for what many regard as Britain's quintessential invasive species: the grey squirrel. As we've seen, the pervasiveness of this North American rodent is wholly due to the activities of Thomas Brocklehurst, Herbrand Russell and other nineteenth-century acclimatisers. The grey squirrel specialises in gnawing off the outer bark to get at the juicy sap-filled phloem, so it's no surprise foresters were among the first to take against the species. Squirrels can kill a tree outright should they completely ring-bark the trunk, but even minor damage exposes it to secondary attack from fungi and other pathogens. At best, the value of timber will be downgraded, with broadleaved varieties such as sycamore, beech, ash and oak most vulnerable, although pines and spruces will also succumb. In addition, conservationists worry about the direct impact on fellow woodland creatures, particularly birds, whose eggs and nestlings represent a ready source of valuable protein for the rodent. For the wider public, though, the displacement of the cherished native red squirrel is the greatest charge levelled against the greys. Once numbering in excess of 3.5 million animals, the red squirrel population has today collapsed to a miserable 140,000, and the grey squirrel is almost certainly a key factor. But, as ever, there's more to the tale.

The red was already in long-term decline before Messrs

Brocklehurst and Russell came on the scene. A specialist of conifer plantations – indeed, until recently, regarded as a virulent pest in its own right – the species had almost vanished from its Scottish heartland by the late eighteenth century due to forest clearances. Red squirrels recovered in the nineteenth century after new pine plantations matured, but they were knocked back again as World Wars I and II prompted further tree-felling. During the early 1900s, many also succumbed to highly contagious diseases, such as coccidiosis – a parasitic infection of the gut – not to mention enthusiastic culling by foresters. Between 1903 and 1946, the Highland Squirrel Club presided over the slaughter of 102,900 red squirrels in the north of Scotland alone, a bounty of fourpence paid per tail, rising to sixpence as numbers dropped. Just as mink were able to occupy a niche vacated by disappearing otters, so grey squirrels were often free to scamper into forests stripped of reds.

There's no doubting, though, that where the two varieties coexist, the non-native gets the upper hand. The greys don't directly attack the reds (an early myth had greys biting the testicles off the red males); it seems subtler than that. For one thing greys are less fussy about food, more readily eating acorns, which reds have a tougher time digesting. Equally significant is the greys' role as resistant carriers of a virus, squirrelpox, which kills reds in droves. These and other attributes – there's also a theory that greys are better problem-solvers – may allow the American species to better exploit a shared habitat, thus surviving in higher densities, particularly in hardwoods. The non-native also lives much longer, nine years versus five or six for a red, so squeezes in more breeding attempts. When grey squirrels turn up, reds vanish within 15 years, often much sooner.

Today, red squirrels have gone from most of England, but

are surviving – thanks to human intervention – on a few islands such as the Isle of Wight, Brownsea Island and Anglesey; the mammal is also dwindling in Scotland and Ireland. Despite Britons' despondency at the loss of their beloved Squirrel Nutkin (more evidence of the influence of childhood reading on our attitudes to wildlife), globally, red squirrels aren't in too much trouble. They're widespread across Eurasia and currently ranked a species of 'Least Concern' on the International Union for Conservation of Nature's Red List. Nevertheless, greys have something of a bridgehead on the continent, having been introduced to broadleaved woodland in Piedmont, northwest Italy, in 1948. As this population was taking off in the late 1990s, Italy's National Wildlife Institute and University of Turin launched an eradication attempt, but court action by animal rights groups stymied this. Grey squirrels are expected to cross the Alps into France and Switzerland any time soon.

Back in Britain, red squirrel devotees aren't giving up without a fight. Everything is being thrown at the greys, which find themselves poked, shot, bludgeoned, drowned and poisoned on an industrial scale. One notable, albeit symbolic, victory has been achieved: on Anglesey, intensive culling since 1998 has removed more than 99 per cent of adult greys from the island within 13 years, using relatively humane methods – live-trapped greys are coaxed into a sack then bashed on the head. This has set the scene for a modest resurgence in reds, some of which, in a twist of fate, benefit from Anglesey's plantations of non-native conifers. Elsewhere the prospects are less rosy. A determined rearguard action is nevertheless being waged, with 16 reserves established across Northumberland and Cumbria set aside for the beleaguered red, among many other such initiatives. In addition, recent pilot studies in Scotland and Ireland suggest that reintroducing or boosting numbers of the pine marten, a native carnivore that preferentially predates greys over reds, supports the recovery of the indigenous squirrel. Hopes also rest on the development of a squirrelpox vaccine for reds, and there's a £1-million government scheme – backed by Prince Charles – to hook greys on Nutella spiked with an oral contraceptive. The doctored spread would be offered in boxes whose lids are too heavy for other small native mammals, including red squirrels, to lift, thus selectively delivering birth control to the unwanted rodents. If all else fails, grey squirrel flesh is said to be delicious. One London eaterie called Native now offers a slow-cooked squirrel ragu, and the rodent meat also turns up alongside rabbit and pork in the 'Critter Fritter', a trademark delicacy dreamt up by Cumbria's Forest Side restaurant. The flavour? Nutty, of course.

The same is probably true of *Glis glis*, another furry trespasser of Britain's forests, which some fear may one day replicate the astonishing success of the grey squirrel. Indeed, as suggested by its common names – the edible or fat dormouse – the culinary value of the species, native to continental Europe and the Middle East, has long been recognised. The ancient Romans kept these rodents in terracotta pots called *dolia* or *gliraria*, fattening them up on walnuts, chestnuts and acorns. The pots would be cooled with water, encouraging their tenants to enter hibernation – a convenient way to store them until it was time for a luxury snack. The dormice would be fished out, roasted, then rolled in honey and poppy seeds.

There's little evidence Romans brought *Glis* to Britain – if they did, it died out soon after their departure. Instead, we have Walter Rothschild, a zoology-obsessed scion of the famous international banking family, to thank. In 1889, for his twenty-first birthday, Walter was given his very own natural history museum in a corner of Tring Park, the family's Hertfordshire estate; well, the empty building anyway. He then dedicated most of his life to cramming it with specimens, many previously unknown to science, including 153 new kinds of insect, 58 birds and 18 mammals. On his death in 1937, Walter left the museum and its contents – 4,470 bird skins, 200,000 eggs, two million butterflies and much more – to the British Museum. It was the greatest single accession that institution has ever enjoyed. But not everything Walter Rothschild collected was dead.

Miriam Rothschild, his biographer and niece, writes of 'kangaroos . . . zebras, wild horses, a tame wolf, wild asses, emus, rheas, cassowaries, wild turkeys, a maraboo [*sic*] stork, cranes, a dingo and her pups, a capybara, pangolins, several species of deer, a flock of kiwis, a spiny anteater, giant tortoises, a monkey . . . and a number of less exotic species' freely roaming Tring Park. For some reason, edible dormice, a handful of which were released at Tring by Walter on 4 February 1902, don't feature in Miriam's list. Perhaps they came under 'less exotic species'. In fact, little is known of the wheres and whyfors of their introduction (one suggestion is that the animals came from Hungary or Switzerland). Yet, *Glis* would prove to be Walter's other enduring gift to the nation, and a far less welcome one.

By the 1930s, edible dormice were breeding in Hertfordshire and surrounding counties, particularly in forests of beech, whose flowers and mast were staple foods; they also turned up as far afield as Shropshire and Coventry – perhaps secreting themselves aboard vehicles. Like the grey squirrels, the rodents have a habit of stripping tree bark to get at the juicy sap within, and, just like the squirrels, soon made an enemy of foresters, particularly those in the Chilterns, whose plantations of larch and Norway spruce came under assault. Before long, householders, from Beaconsfield to Luton, also became acquainted with the mammals, which, at the first signs of autumn, would eschew holes in the ground – their natural overwintering spot – for the luxury of the nearest warm building. The uninvited guests would chew through cables, interfere with roofing felt and ceiling plaster, eat pet food, soil carpets and drown themselves in the toilet.

In a twist of fate, even Walter's own collection is now at risk. As Paul Kitching, the head of the Natural History Museum at Tring, told me, *Glis* now regularly overwinters in the building. 'Every year anywhere between half a dozen and 20 edible dormice are trapped in service ducts under the corridors and lift risers,' says Kitching, who's concerned that the rodent's activity is creating holes in the building envelope. These risk ushering in clothes moths, carpet beetles and other insect pests which could harm the exhibits.

As with the mink and grey squirrel, conservationists are particularly exercised over the dormouse's impact on native wildlife. Once again, there's a suggestion that *Glis* is supplanting a frailer, gentler counterpart, in this case, the hazel dormouse, an animal a fraction of its size.

'The impact on the hazel dormouse hasn't conclusively been proved, but the hazel has been extremely scarce in the Chilterns since the edible dormouse arrived. *Glis* is ten times the size of the hazel dormouse, and can easily kill them.' So said Dr Roger Trout, an ecologist and former Forestry Commission research scientist. Since 2007, Roger and a team of volunteers have been monitoring a population of edible dormice at Hockeridge and Pancake Woods, on the edge of Berkhamsted (some 11 kilometres from Tring). Every spring and summer, they will systematically check around 230 nest boxes sited on trees around the woodland for signs of *Glis*.

When I joined Roger one June, early in the monitoring season, rodents were already emerging from hibernation, leaving their subterranean burrows and heading for the boxes. The latter were each positioned 2.5 metres from the ground: 'It's basically to stop any oiks getting to them,' said Roger. Among the volunteers that day was Anne, a zoology student from Reading University (who carried a ladder); Sean, a local teenager doing the 'volunteering' part of his Duke of Edinburgh award; and Margaret, an older lady who had been working with *Glis* even longer than Roger.

We moved swiftly from box to box through a woodland dominated by mature beech, along with a mix of conifers and other broadleaves. There were healthy stands of bracken and foxglove, with the occasional bumblebee. Each box was probed from ground level using an endoscope – essentially a tiny video camera fitted to a bendy pole. Often, all that could be seen on the monitor were a few freshly plucked leaves, signalling that a dormouse was starting to nest. When the animal itself was present, the ladder would be placed against the tree and the box gingerly lifted from its hooks. The occupant would be dumped into a large polythene bag, weighed and scanned for a microchip, then returned to the box. At this time of the year, the whole process was fairly straightforward. Things got a lot trickier later in the season as the dormice started breeding in earnest, with multiple generations inhabiting a single box. The need to juggle a dozen wriggling – and biting – balls of grey fur wasn't uncommon. A rookie error was to wear shorts and a T-shirt, since any dormouse which fell onto the ground was liable to mistake your legs for tree trunks, and run up them, gripping on with vicious claws.

The first time we struck lucky was with 'Box C5', attached to a red cedar. Inside lurked a 180-gramme male, whose chip scanned as '#966'. This number indicated we were dealing with a two-year-old; they could live up to 14 years. According to Roger, you could tell the sex 'by the gap between the penis and the anus', although I would have thought the penis itself was more of a giveaway. I was struck by just how big these animals were, and how similar to grey squirrels. The main differences seemed to be a slightly less bushy tail, and larger eyes set in a ring of black. 'Given that they're active at night, the ring stops any light reflecting from the fur round and about,' said Roger, who also pointed out the extraordinary length of the whiskers, which were used to detect insects.

Margaret, meanwhile, amassed #966's droppings for clues as to what *Glis* might be eating. The species was already known to kill hole-nesting birds and their eggs, especially tits, nuthatches, treecreepers and woodpeckers, but Roger suspected open-nesters such as blackbirds, thrushes and pigeons were also falling prey. Another line of enquiry was that *Glis* might be taking bats too. Until now, the edible dormice of Hockeridge and Pancake Woods – which today number in the thousands – have been left to their own devices. But, as evidence grows of their ecological impacts, Roger thinks that monitoring will sooner or later turn to 'management', whatever that might involve. Indeed, the Royal Forestry Society, the charitable organisation which owns the woods, is mulling that decision right now. But is it already too late?

9

Freshwater Invaders

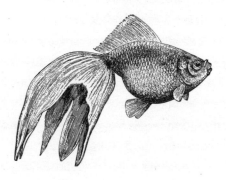

'Thence home and to see my Lady Pen, where my wife and I were shown a fine rarity: of fishes kept in a glass of water, that will live so for ever; and finely marked they are, being foreign.'

The Diary of Samuel Pepys, Sunday 28 May 1665

'If it's a big one, we'll put it back in the river.'

'Fran can castrate it.'

The conversation emanating from the stable-block that morning was jovial. All around were plastic buckets, bin-bags, wooden stakes, wading boots, nets. A fridge-freezer hummed in the corner. The aroma of horse manure and river water hung in the air.

I got speaking to one of the older volunteers, who was stuffing some paraphernalia into a rucksack.

'What have you got there, Tom?'

'Callipers to measure. Knives to kill. Scissors to sterilise. First aid kit. Body bag,' he said. Tom was from Maidenhead but owned a holiday home nearby.

'Site seven today, Tom, OK?' said Nicky Green, who was leading operations. Tom nodded.

Just an hour earlier, Nicky had collected me from Tiverton Parkway railway station in a well-used dark-green Mazda Bongo camper-van. Panting and scampering about in the back was a Lurcher, called Twig, who occasionally rested her muzzle on my right shoulder. While we snaked up the wooded Exe Valley to the farm near Withypool in Exmoor that served as her study base, Nicky spoke of the motivations for her research.

'In Devon, there are only two populations of white-clawed crayfish left, and it's completely gone from southeast England,' she said. 'I was more and more unhappy about the situation and wanted to do something about control.'

In Nicky's mind, the signal crayfish is to blame. The species, originating in the west coast of North America, gets its name from a whiteish patch atop the rear section of the hefty brick-red claws which were once (erroneously) thought to be brandished, flag-like, at other crayfish. Now widespread across England and Wales since its introduction in 1976 by the Ministry of Agriculture, Fisheries and Food to diversify agriculture, signals pass on a pathogenic water mould, called 'crayfish plague', to white-claws, at the same time outcompeting them for food and shelter. (Incidentally, Nicky isn't too concerned by the debate as to the nativeness or otherwise of white-claws, accepting that the crayfish may have been brought to Britain from the continent during the medieval period. As she recently put it in an email: 'Yes, it could be introduced, but do we have a responsibility to assist in the protection of a species endangered to Europe?') Along the way, the signal's burrowing activity undermines river banks, increasing the risk of flooding. The silt which is released into the water smothers invertebrates and clogs the spawning grounds of salmon and brown trout, whose eggs and fry the signal will also consume.

Formerly an ecological consultant, Nicky had been working off and on with signal crayfish for many years, and was all too aware of the difficulties in eradicating them from Britain's waterways. Tactics such as the release of chemicals risked harming other wildlife. Simple trapping wasn't much good either, as this tended to remove only the larger animals; that was a problem because, in natural crayfish populations, the biggest males dominate, monopolising females, preying on juveniles and stopping smaller males from reproducing. Take them out of the system and you create more crayfish than you started with.

But an intriguing alternative – one that Nicky was now trialling on the signal-infested River Barle in Exmoor as part of her PhD research with Bournemouth University and Cefas (the Centre for Environment, Fisheries and Aquaculture Science) – was to trap and sterilise the largest males, then return them to the water. The idea, which had already been tested in France, was that these neutered males would continue hogging the breeding opportunities but fail to leave progeny. Strictly speaking, the crayfish weren't actually sterilised; instead Nicky snipped off two pairs of abdominal appendages, known as pleopods, which male crustaceans used to transfer ropes of sperm to the female during copulation. But it came to the same thing: the next time a tinkered-with male tried mating, his seed would fail to reach its target. (Aware of the potential animal welfare implications of this approach, Nicky argues that crayfish have fairly primitive nervous systems without pain receptors, and points out that they regularly lose and regrow appendages, such as legs and claws, with no ill effects. Indeed, she has recently conducted additional research demonstrating that the behaviour and survival rates of sterilised males are unaffected by the procedure.)

If the approach Nicky is taking succeeds, the implications could be enormous, because when it comes to crayfish, it's no longer just signals we need to worry about. The current tally of non-native crayfish species established in British freshwaters stands at seven, including the narrow-clawed and red swamp crayfish we met in Hampstead Heath. There's also the noble crayfish, introduced to Somerset from central and eastern Europe in the 1980s, and in recent years three further North American kinds have joined the list, or are about to: the spiny-cheek, the virile and the marbled. The last of this trio, also known as the Marmorkrebs and originating in the aquarium trade, is of particular concern, being the only crayfish known to reproduce asexually. All are jostling to dominate their new homes, and many also carry the plague.

By mid-morning we were at the river's edge. Today's complement of volunteers – 13 in total – had been divided into three teams. Each would work a different section of the Barle, which here took the form of an upland spate: narrow, fast-flowing and acidic. These conditions were not ideal for crayfish, which generally prefer waters with higher levels of dissolved nutrients, particularly calcium, which is used in shell-formation (a Hampshire chalk stream would be perfect). Nevertheless, the population of signals was estimated to be thousands-strong in this small stretch of gurgling freshwater. Joining Nicky and me in our group was Ali, from Exmoor National Park Authority, and Louise, a sixth-form student pondering whether to study geography at university. Over the next few hours, we moved upstream – either on the bank, or through the water itself – examining a succession of 50 traps submerged at ten-metre intervals, their locations indicated by wooden stakes.

Two types of trap had been deployed: conical baskets of black plastic, known as baited traps, into which Nicky would sling an opened pouch of cat-food, or an opened tin of sardines; and artificial refuge traps, or ARTs, for short. The latter – the ones we were checking today – resembled giant panpipes, being comprised of eight adjacent plastic tubes of varying diameters fixed to a perforated aluminium base. The ARTs were each weighed down on the river bed with a large rock. If all went to plan, the crayfish would seek shelter in the tubes, believing them to be natural burrows or crevices. Unlike the baited traps, which tended to draw in the larger males, Nicky had found that the ARTs were less biased, catching an equal ratio of males to females and a wider size range.

At each stake, we took turns to wade into the current, pull away the rock and retrieve the ART, many of which were now covered in slimy brown algae. Should a hunkered-down crayfish be present in one or more of the tubes, the trap would be upended and shaken. The contents, which slid into a blue bucket, were sexed, then measured using stainless steel callipers. (The first crayfish I held slipped from my grasp, my middle finger sustaining a sharp nip as if attacked by an angry clothes-peg.) Then came the moment of truth: if the crayfish was a female, she would be killed as a matter of course. The same fate awaited any unfortunate males whose carapace meas-ured less than 40 millimetres in length. The method of dispatch was intended to be as humane as possible, the slippery animal placed on a small wooden chopping board, a scalpel passed swiftly through thorax and head. 'The funny thing is, I'm a vegetarian,' said Nicky. 'A vegetarian murderer.' A '40-plus' male, meanwhile, would be turned upside down, given 'the snip' and thrown back – *sans* pleopods – into the water.

After four years of trapping and sterilisation, Nicky's work on the Barle seems to be paying off, when compared with a different section of the river with no intervention. She has been reassured by a substantial drop in the overall numbers of signals caught per trap, alongside an 85 per cent decrease in the incidence of berried – or egg-bearing – females captured between 2016 and 2018. There's also a noticeable decline in the numbers of one- and two-year-old animals as the effects of the sterilisation come through. But overall catch numbers remain high. Nevertheless, Nicky is optimistic about the approach, should it be applied for the full five or six years of the crayfish lifecycle, and hopes it paves the way for the cradication of the non-native.

Rivers, lakes, ponds, streams, marshes, swamps, bogs, fens, estuaries and other wetlands cover less than 2.5 per cent of the planet's surface, yet they harbour a quarter of what the International Union for Conservation of Nature regards as the world's '100 worst invasive alien species'. That figure doubles when you include the harmful non-native plants, animals and other organisms which occur in terrestrial habitats adjacent to freshwater. The consequences of a new invasive species establishing in an aquatic environment can often be far worse than on land, since such ecosystems are already suffering disproportionately from pollution, habitat modification and fragmentation, climate change and other threats. Among the most frequently cited effects of freshwater invasives is the direct displacement of natives through competition, predation, disease or a combination of all three. They also transform habitats by undermining river banks, muddying the water and blocking watercourses. And every so often, something comes along that changes the way an ecosystem works.

Many of the most troublesome invaders tend to produce vast numbers of usually microscopic propagules: the technical term for their seeds, eggs, spores, larvae or clonal fragments. Therefore, controlling and eradicating unwanted organisms in a complex, three-dimensional system like a freshwater habitat, is often far more difficult than on land. Furthermore, freshwaters are themselves particularly susceptible to colonisation by exotic wildlife, desirable or otherwise. Why? A big part of the answer is gravity.

Wetland ecosystems function like gigantic sinks, with any non-natives – including their propagules – present in higher elevations liable to wash down into them. And, because water bodies tend to be connected to one another via a network of streams, canals, ditches and other conduits, any species reaching one, soon starts turning up everywhere else. Even isolated freshwater habitats, such as mountain lakes or head-water streams, aren't immune. That's because people have a knack of moving aquatic non-natives everywhere they go, either deliberately, for instance when stocking an inland lake with rainbow trout or crayfish, or inadvertently, via contaminated boats, clothing and other equipment. Many of the most intractable species can withstand long periods out of the water, even severe desiccation, so it's easy to see how, with our help, they can leapfrog from one watery spot to the next.

Thanks to its relatively cold climate, Britain has – so far –

avoided some of the worst offenders, including poisonous cane toads, disease-carrying Asian tiger mosquitoes and lake-clogging water hyacinth, but its freshwater environments nevertheless host a growing number of unwelcome introductions. The New Zealand mudsnail is one such example. No larger than a child's fingernail when full-grown, this yellow-shelled mollusc is such a voracious feeder of the microscopic plants and other organic material which support food webs that entire aquatic ecosystems can collapse. The mudsnail tolerates a broad range of aquatic environments, infesting lakes, rivers, reservoirs and estuaries. It also reproduces asexually – a single female is enough to start a colony – so in the right conditions, it forms dense assemblages, hoovering up everything in sight. The species has, in fact, been present in Britain since the middle of the nineteenth century – arriving in drinking water barrels from Australia – and is today widespread. For a while in the 1900s, a burgeoning population blocked London's water pipes, but its current impacts seem modest (although that may be because no one's really checking). Other parts of the world haven't been so fortunate; mudsnail is today found from North America and Scandinavia to Iraq, Russia and Japan. In Yellowstone National Park, USA, some 750,000 were recorded per square metre.

Many of Britain's most problematic freshwater non-natives first arrived for ornamental or aquaculture purposes, before escaping into the wild; others were released into nature right away. At the last count, around 60 invasive freshwater species are known to have established in Great Britain, mostly plants and invertebrates. This figure, probably an underestimate, is expected to rise, partly due to the growing popularity of angling tourism: every time a Brit returns from an overseas fishing holiday, one or more aquatic invaders are likely to hitch a ride back too. The movement of boats and watersports equipment provides a further invasion pathway. In recent years, the rules have begun to tighten. No longer can new species be frivolously released into our waterways, and biosecurity is taken far more seriously. One awareness-raising measure is the government's 'Check Clean Dry' campaign, launched in 2011, which encourages recreational water-users to inspect their equipment and clothing for living organisms, before washing and drying them. Meanwhile, new techniques such as environmental DNA analyses, in which water samples are checked for traces of non-native genetic material, offer a speedier way to detect aquatic interlopers. For instance, the approach is being used to test for the presence of Pacific pink salmon in Scotland's Loch Ness. The scientists involved are keeping an open mind as to whether they might also find monster DNA.

Fish constitute the most obvious exotic addition to waterways, with at least one in four of Britain's freshwater species suspected to be a non-native. As we have seen, the history of fish introductions dates back centuries to the stocking of monastic ponds with common carp for culinary purposes. Fish have also long been collected, traded and bred for their beauty and grace alone, as well as for the status conferred upon their owners. Ornamental fish-keeping of this kind has been practised for millennia in China and elsewhere, but was a more recent trend in Britain, where it was more or less kick-started by the goldfish.

When this small but flamboyant cyprinid (the scientific name for a member of the carp family) first arrived from Asia is unclear. Frequent reference is made to a 1665 entry in the diaries of Samuel Pepys, which describes 'finely marked . . .

foreign' fishes 'kept in a glass of water', but goldfish are strangely absent in later, perhaps more authoritative, works such as Francis Willughby's *De Historia Piscium* (1686) and John Ray's *Synopsis Methodica Avium & Piscium* (1713). Goldfish were certainly collected from China in 1728 by Sir Matthew Decker, director of the East India Company (the same Sir Matthew Decker who would later bring mandarin ducks to Britain); and by the 1750s, the species was known to have been a popular pet in British households.

Like other carp, goldfish are hardy souls, able to eke out a living in the muckiest of waters, and will survive for decades in the wild, where they return to their natural bronze coloration. But goldfish can only breed when water temperatures hit 20 °C, so are unlikely to cause serious problems. There's been a worry that the fish might hybridise with Crucian carp, which are traditionally considered to be native. In fact, the latter may themselves also be introduced. What's more, so similar are the two carps, that some taxonomists suspect them to be different versions of the same species. Perhaps, therefore, the goldfish is nothing to fret about after all. Far greater concern attends another, much more recent piscine introduction to our waterways; one with gruesome eating habits.

The topmouth gudgeon – also known as the stone moroko or false harlequin – is a small, nondescript, silvery-grey cyprinid hailing from the river basins of China, Taiwan, Japan and the Koreas. Yet it's now causing great consternation among the ecologists of Britain. As well as spreading parasites and preying on the eggs and larvae of native fish, topmouth will approach adult fish, stripping away their scales to nibble at the exposed flesh.

The species first appeared in eastern Europe during the

1960s, an unwanted fruit of close cooperation between China and the Soviet Bloc countries keen to develop their own carp-based aquaculture industry. (One theory is that it reached a Romanian fish farm hidden in a delivery of Asian silver carp eggs.) The topmouth flourished in the Danube river system, establishing itself across an average of five new countries every subsequent decade. Then, in the mid-1980s, it turned up in a Hampshire fishery.

How it got here remains a matter of debate. Some believe the fish had hitch-hiked to Britain, others say it was brought here on purpose. The latter hypothesis seems possible: while not much of a looker, the topmouth's upturned maw produces a distinctive crackling noise (for which it earns yet another name, the 'clicker barb'), and importers of ornamental fish may have seen novelty value in this. In the late 1990s, a decade after its initial introduction, topmouth started cropping up in lakes and ponds as far afield as Kent, the Chilterns, Devon and the Lake District. By 2004, they had been confirmed at 23 sites across England and Wales, and were suspected at a further 12.

Why did topmouth gudgeon spread so fast? For a start, their diminutive size – a full-grown adult is a mere ten centimetres in length – lends them a low profile, helping them move around more or less unnoticed. They're even known to hide themselves away inside the mouths of bigger fish. Like others in the carp family, the topmouth is a generalist, able to tolerate a range of habitat conditions, including, for short periods, moderately salty waters. But a number of other characteristics give topmouth gudgeon the edge. For a start, they attain sexual maturity within a year, which is unusual for a cyprinid. In addition, males guard fertilised eggs from predation, boosting hatching success. Topmouth fry will emerge over a long period of time, which allows the species to hedge against unpredictable climatic conditions. All this helps the topmouth to rapidly monopolise a range of new habitats. In Britain, the fish generally turns up in still waters, suggesting it is getting moved around, either as a contaminant of movements of ornamental fish or as a bait fish by anglers. This preference for landlocked water bodies seems to be the topmouth's Achilles heel, since it offers those who wish to remove it the potential for total eradication, although the control method of choice is drastic, to say the least.

The first step is the removal of native fish from the lake or pond to be treated. This is often achieved by electrofishing, a process which involves zapping the water with high voltage and collecting all the stunned fish that float to the surface. Any indigenous varieties are kept safe in dedicated tanks, their mouths and gills checked for any topmouth stowaways. Then, rotenone, a poison naturally occurring in certain plants within the legume family, is applied to the water body to kill off the topmouth. (Rotenone is sometimes described as a 'fish poison', but can also harm a range of invertebrates, and is mildly toxic to humans.) After a few months, the toxin degrades and natives can be safely returned to the water. By 2018, the Environment Agency claimed to have more or less exterminated the topmouth gudgeon, along the way removing two further non-native fish species by way of a bonus. One was the fathead minnow, indigenous to North America and Mexico, which was first recorded in the wild in Britain in 2008 and eradicated two years later. Just as well, since the fish passes on enteric red-mouth disease to trout and eel, a lethal infection, which causes haemorrhaging in the mouth, fins and eyes. The black bullhead, another New World introduction, was second to go; the last specimen of this omnivorous catfish with poisonous spines, introduced during the late nineteenth century, was removed from an Essex fishery in 2014.

The trade in freshwater ornamentals extends beyond fish to other aquatic beasts, including a variety of amphibians and reptiles. Most of these warmth-lovers fail to gain purchase in the wild – we can thank the miserable British weather for that – but occasionally something arrives that seems to make itself at home.

A case in point is the marsh frog. A large and attractive species from central and eastern Europe, the amphibian had been brought to Britain off and on since the late nineteenth century but not naturalised. All that changed in the winter of 1934/5 when Edward Percy Smith, a playwright, and later a Conservative Member of Parliament, released a dozen specimens, collected in Hungary, into his garden at Stone in Oxney in Kent. Smith's pets hopped into the nearby Romney Marsh and, over the succeeding decades, built up a population that today numbers in the tens of thousands. Locally known as laughing frogs, for their thunderous, quack-like calls, marsh frogs are insatiable consumers of all sorts of prey from dragonflies and spiders to small fish, rodents and even birds. Native amphibians also get eaten, although the marsh frog tends to prefer different habitats. Another wannabe colonist, the American bullfrog, hasn't yet replicated the success of the marsh frog in Britain, but has come very close.

Bullfrogs are now causing havoc in more than 40 other countries, where they are implicated in the transmission of fatal chytridiomycosis to indigenous amphibians. The species has primarily been shipped around the world for cultivation, its chunky hind-legs a much-loved delicacy, but the pet trade served as the primary invasion pathway into Britain. At first, bullfrog tadpoles – which measure up to 15 centimetres in length – would contaminate shipments of ornamental freshwater fish from North America, but then Brits took a shine to the amphibian as a pet in its own right. A vibrant bullfrog trade built up, until the New World croakers started turning up in the wild. The authorities stepped in, and imports were banned from 1997. But by then it was too late.

Just two years later, the species was found to be breeding at seven ponds in Edenbridge, on the East Sussex–Kent border. An intensive culling campaign against the bullfrog was launched, with air-rifles, shotguns, catapults and pitfall traps among the weapons deployed. Two ponds had to be drained before the population was at last wiped out in 2004. In the space of just five years, no fewer than 11,830 bullfrogs were removed at a cost of £100,000. A couple of years later, a new population was nipped in the bud in Essex. 'The battle against the bullfrog' has similarly been waged in France, Germany, the Netherlands and Italy. In Belgium, which also has several bullfrog populations – including a serious outbreak in the Flanders region with more than 400 affected ponds – scientists are thinking about borrowing a technique from the signal-crayfish playbook: the release of sterile male bullfrogs. The latter, created by applying a high-pressure shock to the frog embryos at the moment of conception, are soon to be released into the wild in the hope that they'll outcompete their fertile counterparts.

A very different class of non-natives is now making incursions into the freshwater realm, organisms that are proving far more of a headache than crayfish, topmouth gudgeon and bullfrogs.

Water fern. Floating pennywort. Parrot's feather. Water primrose. New Zealand pigmyweed. They sound innocuous, even whimsical, but the impacts of these aquatic plants – many of them prized ornamentals – have been judged so severe that all five are now banned from Britain; anyone selling them risks a £5,000 fine or six months in jail. Like their terrestrial counterparts, these and other invasive water plants spread fast – often reproducing asexually – smothering the competition, messing up habitats and spoiling the fun for recreational water-users. But tackling plants in an aqueous environment is far tougher than on land – spraying them with herbicide risks harming native species and when they're in deeper water, getting close can be dangerous and difficult.

Among the worst is New Zealand pigmyweed, also called Australian swamp stonecrop, common names that point to its Antipodean origin. Identifiable by its tiny succulent leaves, pinkish-white flowers and long straggly stems, this aquatic perennial made its debut in 1911, when some plants were brought over from Tasmania to oxygenate British ponds. This is something of an irony, since pigmyweed often does the reverse, forming dense floating mats of vegetation and slowing down currents that would otherwise aerate the water. In addition, when the plants die off, the rotting process further depletes oxygen from the system, killing off fish and other aquatic organisms. As with many invasives, the species took a while to naturalise; the first wild record wasn't until 1956, when the plant was spotted in ponds near the Essex village of Greenstead. Since then, pigmyweed has made up for lost time and is today found throughout much of Britain and the European continent, as well as in the USA. Getting rid of it is almost impossible.

That's not for want of trying. In the New Forest, which has had a pigmyweed problem since 1976, they've thrown everything at it, including Roundup herbicides, boiling hot foaming agents, and dye treatments to stop photosynthesis.

Nothing seems to work. Similar efforts have been undertaken in the Broads of East Anglia, where they've also tried covering the stuff with black plastic bags and have even resorted to digging a whole new pond, transferring everything across – minus the pigmyweed – and burying the old one. And now, new research suggests the pigmyweed threat could be even worse than feared.

Until recently, the species was assumed to be dispersing itself solely through vegetative reproduction, with minuscule fragments of stem carried along on water currents or conveyed to new locations by wildlife and human activity. That was bad enough. But in 2014, greenhouse tests showed that pigmyweed was also producing viable seeds and may have been doing so all along. This is forcing a rethink on control: it's now no longer enough to remove all above-ground vegetation, because the seeds – almost invisible to the naked eye – will probably help the plant bounce back.

Floating pennywort, another aquatic plant on the banned list, is a more recent arrival whose effects on freshwater systems – dense mats of vegetation, deoxygenation, and so on – are similar to those of pigmyweed. This one came from North America, via the horticulture trade, in the 1980s, and got going in Britain unnoticed due to its close resemblance to the indigenous marsh pennywort. The non-native species is said to spread at up to 20 centimetres per day, and by 2017 was recorded at more than 1,000 sites. In October of that year, huge rafts of pennywort appeared on the Rivers Cam and Great Ouse in Cambridge, threatening to interfere with the university's rowers and punters. A £30,000 emergency eradication effort was mounted with metres-long mats of thick impenetrable vegetation, some weighing almost two tonnes, hauled from the water.

The Cam is no stranger to such episodes. In the nineteenth century, the same stretch of river was clogged up with Canada waterweed. The New World invader – then called 'anacharis' – was first identified in the autumn of 1842 in the lake of Duns Castle in Berwickshire. Around the same time, it also appeared at the Union Canal's Foxton Locks in Leicestershire, from where, in 1847, a specimen was collected for the newly opened Cambridge Botanical Garden. The curator, Andrew Murray, then made the mistake of placing a sprig of the plant in a nearby stream. A contemporary account in *The Leisure Hour* describes what happened next: 'After spreading all over this site, it seems to have bolted through the waste pipe, across the Trumpington road, into the Vicar's brook, and from thence into the river, where it was soon conspicuous.' The Cam was rendered almost impassable: extra horses were needed to drag barges through the mass of vegetation, rowing was impossible and bathers got tangled up, one or two reputedly drowning. Similar scenes played out in rivers, streams and canals across Britain, with the outbreak peaking during the 1860s, before the plant mysteriously declined in the early twentieth century. While dramatic boom-and-busts like this are considered characteristic of invasive species, they are rarely proven to occur, and those now dealing with rampant pigmyweed and pennywort must be praying that history repeats itself.

Every day, a quarter of a million cars, trucks, vans, buses and bikes speed (or more often crawl) between junctions 14 and 15 of the M25 motorway. This three-kilometre stretch of ten-lane tarmac west of London serves the country's most densely populated corner, and Heathrow, its busiest airport. It's among the most clogged highways in Britain, if not Europe. But how many of the drivers and passengers sitting there, sipping tepid lattes, checking their smartphones or marvelling at the succession of airliners in slow descent, realise that a mere stone's-throw from their vehicles, congestion of a very different kind is also building up, one that threatens profound changes to Britain's freshwater habitats?

That's what Daniel Mills fears. He's a PhD student from the geography department of King's College London and has been documenting the impacts of a brand-new addition to our aquatic fauna, an organism that he fears could multiply out of control. Multiply out of control: this is a tired phrase in the context of invasive species, but when it comes to the quagga mussel, it's an eminently reasonable description.

One sultry late summer morning, I accompanied Daniel to his study site on the River Wraysbury, close to the Surrey–Berkshire border. This stretch of water, a place he dubs 'Ground Zero for the quagga', was best accessed on foot from nearby Staines across an expanse of scrubby land known – perhaps inevitably – as Staines Moor. Joining us was a fellow geography student, Eleanore Heasley. Daniel had led the way, lugging a hefty backpack from which protruded a length of wooden ruler. In one hand he carried a strange piece of aluminium apparatus. Several times we crossed, by low bridge, the meandering Colne river, of which the Wraysbury was a tributary. In the middle distance, a relay of Boeings and Airbuses could be seen, the occasional furious rumble carried our way on the changeable wind. We passed horses, cows and numerous ant-hills, before reaching the Wraysbury itself.

'River' was perhaps too generous a title for what amounted to a slender channel of water tucked away below the aforementioned section of motorway. The smell of exhaust fumes and burnt clutch wafted downward; the drone of engines was constant. Judging from its geometrical straightness, I supposed that the Wraysbury's natural kinks had been ironed out to accommodate the road. The iridescent, fast-flowing water nevertheless retained a certain beauty. Electric-blue damselflies darted at the surface, while beneath it trailed the vivid green fronds of water buttercup, harbouring shoals of little bullhead fish. Eleanore and I donned thigh-high wading boots. Daniel, who was already wearing his, stepped down into the stream. 'Look at the density of them!' he cried. 'No wonder they're called "ecosystem engineers". They settle here, growing in dense beds and binding pebbles together. They're completely altering the nature of the river bed itself.'

The knee-deep water offering a clear view of the bottom, I could see what he meant: almost every stone was crusted with the tiny striped shells distinctive of the quagga mussel (the animal was so named due to the similarity of its pattern to an extinct type of zebra). These freshwater bivalve molluscs, seldom more than four centimetres in length, were notorious for latching on to any available hard surface with tough strands of keratinous protein. Then, when the mussels died a couple of years later, new ones settled on top. If they weren't scraped off, quagga shell layers accumulated in nature's answer to the Tetris video game. In other countries, where the species has established, pipes have been clogged, boats fouled and other varieties of mayhem have ensued.

Since the late twentieth century, ecologists and engineers alike had been watching in dismay as the quagga mussel progressed westwards from its native Ukraine, specifically from the estuaries of the Southern Bug and Pripyat, major rivers which emptied into the northern Black Sea. The adult mollusc was shown to disperse itself in numerous ways, from hitch-hiking on drifts of aquatic vegetation to sticking to boat hulls. It's even been known to catch a ride on the hard exoskeleton of a passing crayfish. The quagga mussel's planktonic larvae, called veligers, were more mobile still, able to drift on water currents in their own right. A single female quagga produces close to a million veligers a year, so if you spotted mussels in the headwaters of a new river, you could expect, sooner or later, to see them appear downstream as well. If that wasn't enough, countless veligers were also sucked into the ballast tanks of river and even seagoing vessels, then spewed out in distant locations when the water was expelled. Ballast transportation best explains how quaggas appeared in the Great Lakes of North America as early as the 1990s and reached the canals of the Netherlands by 2006. It was a matter of time before Britain too would have a 'quagga problem'.

Sure enough, in October 2014 the species was in the Wraysbury. No one is certain how quaggas arrived, but Daniel suspects that they may have been first introduced to a nearby reservoir, and from there got into the river. A statutory obligation on the water company every five years to test the reservoir's emergency valve may have unwittingly resulted in the quagga's breakout: 'Even if they'd only opened the valve and run out some water from the reservoir for a few minutes, that might have been enough for some of the mussel's veliger larvae to escape.' Since then, it has spread down interconnected watercourses. For example, within three years quaggas were turning up in the Thames at Richmond, some 30 kilometres from their initial discovery. Meanwhile, back at Daniel's study site, if you were to scoop up and weigh every invertebrate from the river bottom, the mussels would represent 60 per cent of the total biomass. Other species in the Wraysbury were beginning to feel the consequences, none more so than the indigenous unionid mussels, a group of larger, longer-lived and more solitary bivalve molluscs.

Unionids – which include 'painter's mussels', so called because artists used to store pigments in their close-fitting shells – need a sandy substrate in which to bury themselves. But with almost every square centimetre of river bed carpeted with quaggas, the much larger unionids were fast running out of suitable real estate. Adding insult to injury was that quaggas also colonised the shells of the natives themselves, forming miniature molluscan minarets. The invaders consumed microscopic algae, filtered from the water column with awesome efficiency as their population densities grew. Unionids depended on the same stuff so, given the sheer weight of quagga numbers, they were being starved, as well as smothered, by the non-native. Less visible but more far-reaching impacts were also possible.

'It's part of a chain,' explained Daniel. 'The algae consumed by the quaggas are also eaten by zooplankton – *Daphnia*, those sorts of things – which are, in turn, eaten by the juvenile fish, and juvenile fish are then eaten by the larger fish. So, algae are the primary producers in the water; if you knock them out, the effect on the entire ecosystem can be significant.'

The quagga is just the latest in a small army of aquatic animals from the region of the Black, Azov and Caspian Seas that has been overrunning waterways across western and northern Europe in recent decades. Collectively known as Ponto-Caspian species, these seem to have evolved in rivers, estuaries and lagoons, conferring on them a natural tolerance to wide fluctuations in water salinity. This adaptation has in turn enabled them to exploit a broad range of new habitats. However, not all Ponto-Caspian species are newcomers.

The zebra mussel – a close relative of the quagga with similar impacts, distinguished by a flatter bottom shell edge – reached Britain as early as the 1820s, aboard timber ships from the Baltic, and dispersed through the canal network. Their effects have worsened in recent times: in 2011, almost 800 tonnes of zebra mussels had to be removed from a tunnel into the Walthamstow waterworks in east London, only to build up again a few years later. (They cause even more problems in the Great Lakes region of North America, which they reached in 1988, courtesy of the ballast tanks of transatlantic freighters.)

The trickle of Ponto-Caspian invaders has lately become a flood, attributed to the construction of major new canals, such as the 170-kilometre-long Rhine–Main–Danube, completed in 1992, and the renovation of older ones, like the Mariinsk, which links the Volga river to the Baltic Sea. These artificial watercourses are acting as new trans-continental highways for invasion, with aquatic organisms hitching a ride on passing ships or simply drifting with the current. Since 2000, at least five new Ponto-Caspian fish species have turned up in the Netherlands thanks to the canals, including the round goby, which has been shown to displace natives. Sometimes the speed of dispersal has been astonishing. In the Rhine river catchment, six types of freshwater mollusc or crustacean were found to be spreading at an average annual rate of between 44 and 112 kilometres. One species, the killer shrimp, a fast-breeding omnivore notorious for indiscriminately shredding invertebrate prey with outsized mouthparts, covered an astonishing 500 kilometres in a single year. Killer shrimps turned up in Grafham Water, Cambridgeshire, in 2010. The roll-call of recent Ponto-Caspian introductions includes other crustaceans such as the demon shrimp – a close relative of the killer shrimp – first reported in 2012 in the River Severn at Tewkesbury, and the bloody-red mysid, discovered in Nottinghamshire's Erewash Canal in 2004.

Many more species, entrenched in waterways just across the Channel, are poised to invade.

Not every significant freshwater invader exploiting Europe's new aquatic highways originates in the Ponto-Caspian region. The Asian clam, for example, is native to China, Korea, Japan, Thailand, the Philippines and southeast Russia – where it is widely eaten – yet once it got into the Rhine in the 1980s, it quickly spread across the rest of Europe via the interconnected water systems. It has also been introduced to North America – where it gives the zebra mussel a run for its money – as well as South America and Australia. The species is thought to have arrived in Britain in 1998 via ballast water (although the clam was also allegedly introduced to Ireland for culinary reasons) and has since proliferated in the Broads of East Anglia, wreaking the same sort of impacts as zebra and quagga mussels. Unusually for an aquatic invader, the Asian clam is rather sensitive to pollution, preferring well-oxygenated waters, so recent new regulations to improve water quality may inadvertently be playing into its hands. One new control technique now being mulled over in Ireland to deal with the Asian clam is thermally shocking the mollusc with dry ice pellets; it seems to work in the lab but hasn't yet been tried in the field.

Most alarming of all is that Ponto-Caspian species seem to be facilitating each other's invasiveness in a vicious cycle which experts call 'invasional meltdown'. For instance, the jagged topology of the river bed, caused by the accumulation of quagga and zebra mussels, offers perfect hiding places for demon and killer shrimps, whose co-evolved striped patterns allow them to blend in. The shrimps – which benefit from a ready source of nutrition, the mussels' pseudofaeces – are in turn eaten by other larger non-natives, including fish like the round goby, whose population is now massing in the rivers of western France. These interactions are already occurring on the continent, and recently the demon shrimp has also appeared in the River Wraysbury, suggesting the pattern is repeating itself here as well.

There was no mistaking Daniel's attitude to these invaders from the East: 'Because of us, these animals have circumnavigated 50 million years of evolution, give or take. They've come in and they've cheated, *cheated*, effectively. The conditions here are much cushier than back in their native lands, and as an ecologist, that perturbance of the natural system, that change, I don't know if it should sit easily with people.'

Before leaving the Wraysbury, Daniel revealed some intriguing news: something was taking quagga mussels from the water, drying them on the bank and consuming them. The evidence took the form of mysterious piles of shells. 'That sort of behaviour seems mink-ish or otter-ish to me,' he said. The plan was to install camera-traps to solve the mystery. 'If it is a mink, then it would be a further case of one invasive species helping another to proliferate!' It seems the full story of the quagga – and the rest of Britain's freshwater invaders for that matter – was set to run and run.

10

Underneath the Waves

*'Time and tide wait for no man. A pompous and self-satisfied
proverb, and was true for a billion years; but in our day of
electric wires and water-ballast we turn it around: Man waits
not for time nor tide.'*

More Tramps Abroad, Mark Twain, 1897

Brighton. June 2015
A trio of charter boats pushed off from Brighton Marina. The
mood of those aboard – 120 people in flamboyant attire – was
boisterous but reverent. A kilometre or so off the Sussex coast,
the engines were cut and live crabs and lobsters dropped into
the sea. The animals weren't simply flung over the side: the
people first blessed each in turn, uttering prayers and dedica-

tions to deceased loved ones. Once destined for the pot, the shellfish – there were close to a thousand of them, costing a total of some £5,000 – would now enjoy a new lease of life. In return, the human participants, through their demonstration of empathy for the natural world, would receive absolution for past misdeeds.

The English Channel had just hosted the Buddhist ritual of life release, or *fangsheng*, a practice dating back more than a millennium whose popularity has resurged in recent years. Good karma all round, but for a wrinkle in the plan. Hundreds of the crustaceans tasting freedom that day weren't native.

This soon became all too apparent, when a local fisherman, bewildered by the unusual new crabs and lobsters he was catching, alerted the Marine Management Organisation. An arm of the UK government charged with protecting the seas, the MMO took a dim view. A full-scale recovery operation was launched with the assistance of Cefas, with boats chartered at an eventual cost of £18,000, and a £20 bounty offered for every non-native crustacean recaptured. One of the experts involved later described the episode as complete mayhem: 'I was absolutely, ridiculously busy . . . long, long hours . . . had to brief ministers and get evidence to fund the control effort.'

A subsequent investigation found the Brighton ceremony had been planned by a pair of London-based Buddhists in their thirties. Although some animals released that day were native lobsters and spider crabs procured at nearby Shoreham harbour, the numbers had been augmented by 361 American lobsters and 350 Dungeness crabs (a Pacific species), supplied by a Greenwich-based specialist fish wholesaler. The ritual had been inspired by a visit to Britain of the Venerable Hai Tao, the Taiwanese founder of the Compassion for Life Organisation. (It turns out that Hai Tao, who heads his own television channel and is an enthusiastic proponent of *fang-sheng*, has form: three years before this episode, his disciples liberated hundreds of exotic snakes into the hills of Taiwan.) In September 2017, the two Buddhist organisers pleaded guilty at Brighton Magistrates' Court to contravening the Wildlife and Countryside Act 1981 by releasing exotic species into the wild, and were ordered to pay more than £28,000 in fines and compensation. To date, fewer than half the non-native animals have been caught. No one knows how many are now thriving in the English Channel thanks to this unexpected act of kindness.

Incursions of American crustaceans had already been troubling experts before the Brighton incident. Lobsters were a particular concern: about a hundred had been recorded in European waters over the previous decade, a quarter in British waters alone. Spiritual practices need not be ascribed to all such appearances: as with other live imports, non-native lobsters have had plenty of opportunities to win their liberty. These include cruise ships dumping not-quite-dead animals, accidental releases during transport and escapes from marine holding facilities. With 20 million live lobsters air-freighted into Europe from northeastern America every year, that more haven't absconded is miraculous. But even if they did, would it be a problem? Scientists think so.

For starters, American lobsters grow larger, and are more aggressive, adaptable and fecund than their European counterparts, so threaten to outcompete them for habitats, as well as displacing other native crustaceans such as brown crabs and langoustines. They also carry a nasty bacterial infection called gaffkaemia; commonly known as 'red tail disease', the condition causes captive native lobsters to keel over and die, often shedding limbs in the process, and may impact wild populations as well.

Another concern is hybridisation. In Scandinavian waters, it's been discovered that the two species interbreed. Should the hybrids prove sterile then, at best, a breeding opportunity for the European lobster is squandered. A more troubling scenario has the crosses surviving long enough to themselves reproduce, potentially even displacing the European lobster altogether, a form of extinction. So worried are the Swedes – who have a thriving lobster industry of their own – that in 2016 they (unsuccessfully) petitioned the European Union to ban imports of the American variety. That one of the Brighton lobsters was recaptured with eggs fertilised by a native male, suggests that cross-breeding may have occurred here too, which only heightens concerns.

So, once again, we have the prospect of a tough, new, disease-ridden and oversexed outsider coming in and displacing a feebler native. It's a pattern which should, by now, be all too familiar given our review of non-natives in terrestrial and freshwater habitats. But there's an essential difference: once an unwanted species establishes in a marine environment, getting rid of it is next to impossible. That, of course, doesn't stop people trying.

Take the carpet sea squirt, a colonial invertebrate native to the northwest Pacific, which arrived in France in 1998. Ten years later, it was recorded in Anglesey's Holyhead harbour, probably having spread across from Ireland, and has now turned up in at least nine marinas from the Solent to the Clyde. Sac-like creatures, generally attached to the bottom, sea squirts are so named for the paired siphons used for sucking in plankton particles and expelling waste. The variety in question here forms vast pale-cream or orange aggregations that can smother shellfish beds and fish spawning habitat, or hang from jetties, hulls, mooring chains and buoys in gloopy leathery structures variously likened to dripping candle wax, snot or vomit. In Anglesey, where the local mussel industry feels threatened, a fortune has been spent trying to get rid of carpet sea squirt colonies, which have been scraped, suffocated in heavy-duty plastic bags and treated with bleach. Elsewhere they've tried concentrated vinegar, lime, fresh water and even swimming pool chlorine. But while isolated colonies can be killed, small numbers of 'parent' stock might be growing nearby just out of reach, and when summer comes the sea squirts regenerate themselves like a scene from Walt Disney's *Fantasia*. With marine invaders, the emphasis therefore has to be on prevention. That's why those who release troublesome new species into the sea, even with the best of intentions, risk prosecution.

Notwithstanding the odd religious ceremony, few non-natives are deliberately loosed into the sea in the way that so many plants and animals have been released on land. Indeed, as we will see, inadvertent transport on ship hulls, or via their water ballast systems, are the most important invasion pathways for aquatic non-natives. Nevertheless, a significant number of seagoing organisms are purposely moved around the world in the interests of marine aquaculture – or mari-culture. Thanks to the rearing of mussels, oysters, scallops, lobsters, crabs, trout, salmon and other farmed seafood, at least 19 exotic species are known to have established in Britain's saltwaters; and this is a conservative figure. Although the cultivated organism is itself sometimes the problem, more often than not it's other hangers-on that cause upset. Just as countless terrestrial and freshwater organisms travel the world as hitch-hikers of the horticulture trade, so a panoply of marine life moves between mariculture sites on fouled ropes, chains and other equipment, or as contaminants of imported stock.

Among the best-known farmed species welcomed to our waters in recent years is the Pacific oyster, which today accounts for around 85 per cent of all oysters cultivated in Britain. Commercial exploitation of its native counterpart, the flat oyster, dates back at least to Roman times, with the oysterbeds in Kent's Whitstable Bay and the Blackwater estuary in Essex later enjoying royal protection. The mid-nineteenth century saw a boom in the oyster industry linked to the development of railway lines, which enabled rapid distribution to market of these perishable delicacies. By the late 1860s, some 1.5 billion flat oysters were consumed annually in Britain. But a combination of over-harvesting and river dredging triggered a catastrophic decline, with a series of cruel winters during the first half of the twentieth century delivering the *coup de grâce*. The search was now on for an alternative.

The Pacific oyster, a species native to Japan, China and other parts of northeast Asia, was once assumed to have been a recent introduction to British waters, not appearing until the 1960s. However, as early as 1890, an animal known as the 'Portuguese oyster' was being fattened up off the Dorset coast by the Poole Oyster Company. This variety, already cultivated on the Iberian Peninsula and long assumed to be European, in fact originated in the Far East and had been introduced to Lisbon harbour by sixteenth-century traders, possibly attached to their ships' hulls. Indeed, the consensus is that Portuguese and Pacific oysters are one and the same (although some still disagree). Whatever the truth, the Poole enterprise, and similar experiments with Portuguese oysters elsewhere in southeast England over ensuing decades, failed to take, stymied by pollution and disease outbreaks.

In 1964, the industry was revived with the sourcing of fresh, parasite-free, stock from Canada, where Pacific oysters had been cultivated since the beginning of the twentieth century. With UK government support, the new oysters were propagated in quarantine facilities at Conwy, North Wales, before successful trials across the country. A few years later, France imported enormous quantities of the Pacific species, including 562 tonnes of adult oysters from Canada and five billion seed oysters direct from Japan. The introductions soon formed self-sustaining populations off the French coast. Experts assumed that the colder British waters would prevent this happening further north. The experts were wrong.

That small numbers of free-living 'Portuguese oysters' had on occasion appeared in the Blackwater estuary between 1901 and 1962 was a warning unheeded. From the 1970s, natural aggregations of Pacific oyster larvae, called spatfalls, began to be recorded in Dorset and Scotland. Warmer summers in 1989 and 1990 favoured further settlement in the estuaries of southwest England and the Menai Strait between North Wales and Anglesey. Today, the species is well established off southern Britain, sustaining commercial wild harvesting in the Mersea region of Essex, with the River Yealm in Devon and the loughs of Northern Ireland also hotspots. In certain locations, great reefs of oyster shells are now forming with 200 or more individuals per square metre, reminiscent of the quagga and zebra mussel aggregations that plague freshwaters. Some new colonies aren't close to oyster farms, suggesting Pacific oysters are also being moved around by ships, in their ballast or on hulls. In addition, eggs and larvae drifting across the Channel from the vast oysterbeds of Brittany may found new populations.

Oyster hatcheries around the world have scrambled to

contain the spread by producing triploids: animals with an extra set of chromosomes which, in theory, are sterile. It's not an entirely altruistic act. Regular oysters use up energy reserves during the summer months to produce spawn – a single individual releases up to 100 million eggs – becoming emaciated and watery in the process; doctored ones, by contrast, have no such impulses, remaining chunky and more valuable all year round. In Britain, where triploids are created by applying special chemicals to the eggs at the point of fertilisation, the approach hasn't been a roaring success. For instance, triploids introduced by the Duchy of Cornwall for farming at Port Navas on the Helford river are alleged to have escaped and begun reproducing. (The bullfrog-breeders of Belgium should perhaps take note.)

Deeming as inadequate the official response to Britain's rising tide of Pacific oysters, some environmentalists have resorted to legal action to shut down oyster farms, including the Helford operation (the Duchy emerged victorious but closed it anyway in 2017). Meanwhile, volunteers armed with hammers and power-drills go out at low tide to have a crack at the shells. All this smacks of desperation: in truth, it's too late. The Pacific oyster, already one of the better-travelled marine molluscs on the planet, seems to be here for the long haul. And, beyond its reef-building activities, which alter the physical structure of ecosystems – and the real risk in some places, such as the Yealm estuary in Devon, of people and dogs cutting their feet on the sharp-edged shells – you would be hard-pressed to pin any serious charge on it. Indeed, some conservationists believe the oyster reefs boost biodiversity by providing more complex habitats. And more than that, the animals' constant filter-feeding purifies the water, absorbing dissolved nitrogen and carbon. In fact, it's the plethora of organisms tagging along wherever Pacific oysters are introduced that excites more justifiable concern.

Most of these oyster farm hitch-hikers also originate from the Far East, such as the Asian oyster drill. The name says it all: this predatory marine snail uses rasp-like mouthparts and acidic secretions to excise perfect circles from the shells of prey, before sucking out the soft innards. The species is now well established off the coasts of France, the Netherlands and Denmark, but at the time of writing hadn't yet shown up in British waters. That's just as well, because oyster drills aren't fussy, attacking indigenous and introduced shellfish alike. Oysters also pass on their own coterie of parasites and pathogens. These include an assortment of microscopic crustaceans called copepods, some specialised varieties of which eke out a living in their hosts' guts or gills. There are some fatal diseases too, with names such as MSX, Dermo and bonamiosis; they're caused by mysterious single-celled organisms which occasionally crop up in shellfisheries to disastrous effect. The worldwide oyster industry has also done its bit to spread a weird and wonderful collection of macroalgae: that's seaweed to you and me.

Notable among these is the sea potato, a native of the Pacific which appeared in French waters in 1906, crossing the Channel soon after. Its other name, the oyster thief, comes from the fact that this hollow species will sometimes fill with air and float off at high tide like a cluster of small beige beach-balls. If the weed happens to be stuck to small shellfish at the time, they'll go with it. Entire oysterbeds have been spirited away in this fashion.

Wireweed is another fellow traveller – again from the Far East – which was discovered in Bembridge lagoon on the Isle of Wight in 1971, having already fouled oyster farms in North

America and France. This fast-breeding species, also called smut (and, less diplomatically, 'Japweed'), forms large mats of olive-brown vegetation which break free from holdfasts to tangle up fishing gear, interfere with boat traffic and cause a nuisance to recreational water users. It's a bit like the sea's version of floating pennywort or Canada waterweed. Concerned that the wireweed might smother the fragile eelgrass communities that shelter cuttlefish and rare seahorses, conservationists had a go at eradicating the Bembridge colony by hand. The effort proved futile, as have subsequent attempts using trawling, cutting and suction. The trouble is that, like many terrestrial invasive plants, wireweed can propagate itself from just a fragment of holdfast, so it's close to impossible to get rid of an established colony. What's more, the removal process itself risks further damaging shorelines. (Fortunately, the Bembridge eelgrass beds survived.) Wireweed has now spread around Britain and Ireland, as well as penetrating into Scandinavia and the Mediterranean.

Professor Jason Hall-Spencer, a marine biologist from Plymouth University, works with a more recently arrived macroalgal associate of the Pacific oyster trade. Japanese kelp, also known as wakame, is colonising coastlines from Mexico and the Mediterranean to Australia and New Zealand, where it's accused of outcompeting native kelps for space and light. The seaweed's advance is not always accidental. A traditional ingredient of miso soup – its flavour is often (unhelpfully) described as 'delicate' and 'salty' – wakame is grown as a mariculture delicacy in its own right; in 1984, three-metre-long specimens were transplanted from the south of France to Brittany for cultivation on ropes. A decade later, wakame hopped the Channel.

But Jason, who studies the weed here and in Japanese waters, argues that its arrival is not a complete disaster. I met him at a marina in Plymouth, Devon, which, like most others on the coast, is a magnet for cosmopolitan sealife. The tide was up, and the greenish soup-like water teemed with mullet and seabass. Drifting on the current was a squadron of compass jellyfish, so named for forks of reddish-brown pigment radiating from the top of the bell. A drowned hedgehog, its spikes just breaking the surface, added to the species count.

Jason led me along a pontoon, stopping to chat with a boatload of diving students in dry suits fiddling with tanks, masks and gauges, their vessel's noisy engine spluttering bilge fumes. He knelt to haul from the brine a collection of chains festooned in glutinous grey, orange and green lifeforms. It had all the appearance of a Christmas decoration gone wrong.

'A lot of what you see here is invasive,' he said. 'These tufts are brown bryozoans, this slimy-looking tunicate and this orange tunicate, they're all invasive species which have arrived since 1957.' He knew that because none are mentioned in a comprehensive study of the local marine fauna written that year.

We moved on to a nearby rope, also trailing in the water; attached were slimy lengths of native and Japanese kelp whose fused holdfasts had, by chance, brought them together in an awkward embrace. Jason laid the olive-green fronds side by side on the pontoon.

'At first sight, they're quite similar,' he said. 'But I can tell that this one is the wakame, because it's got this wavy section at the bottom, and the European one doesn't.'

'What's the function of the wavy bit?' I asked.

'It's where the spores are produced and is the tastiest part.

You can eat the whole thing, but that's the really nice bit. As it gets longer and tattier, organisms cover it.'

And that, for Jason, was the point: should the Japanese kelp one day supplant its European counterpart, many of the same ecosystem services would be furnished, and perhaps a similar richness of biodiversity sustained. A study in another English harbour found wakame supporting a variety of native animals, with 180 sea slugs from 20 or more species on just six strands of the weed. 'This is still providing habitat,' he said. 'It's still taking nutrients out of the water, still cleaning the harbour.'

Jason's ideas may be put to the test soon, as the seaweed is on the move. Many aquatic invasives introduced to marinas tend to stick around, quite literally: preferring to settle on the undersides of harbour structures and moored vessels. But wakame, first recorded in the Hamble estuary, near Southampton, in 1994, wasted no time in venturing out, and has since settled along much of England's south coast, reaching Plymouth in 2003. As Jason put it, 'It's like rabbits: eventually they gain resilience to the local conditions and they get out.'

The oyster trade doesn't have to answer for everything in the marina. The comings and goings of naval vessels are part of the story too. Among oft-encountered non-natives in the marina is a little peanut-shaped bristleworm from the Far East, which spends its life buried head-down in the mud. As Jason explained, the worm probably reached Plymouth sometime in the 1950s in the ballast tanks of American warships returning from the Korean War.

In truth, wakame, oysters, Asian seaworms and the rest didn't have much adjusting to do on arrival. Our temperate waters resemble those of their home range in northeast Asia and, given the volume of maritime traffic between the regions in recent decades, that an increasing number of organisms from the seas around China, the Koreas and Japan are establishing themselves here doesn't surprise Jason. But what does puzzle him is the *one-sidedness* of the exchange.

Just as more species have colonised the New World from the Old than vice-versa, so Asian organisms moving into our waters seemingly far outnumber those going the other way. This might be an artefact of recording effort: the Plymouth estuary and surrounding waters are among the best studied in the world, so perhaps many more European marine organisms lurk in Asian seas than scientists realise. But if the observed East–West imbalance is real, then perhaps this lop-sided exchange is due to the almost uni-directional movement of oysters from Japan to Europe, directly, or via intermediate staging posts. Alternatively, British waters may simply offer more empty niches for exploitation. 'I'm thinking off the cuff,' said Jason, 'but I sort of wonder whether it links to our glacial history. Most of the UK coastline was ice-bound. Now, because of the Gulf Stream, that's all been opened up only in the last 20,000 years. So we've got space for new organisms. There's more opportunity for invasion.'

There's no denying that our weakness for seafood has played a significant role in moving non-native organisms around the world, but as an invasion pathway, mariculture is dwarfed by shipping. With 90 per cent of our commodities today conveyed by more than 50,000 merchant vessels, each capable of harbouring a multitude of species on their hulls or in ballast water, the opportunity for new introductions is enormous and unprecedented. And, as the speed and size of vessels have increased exponentially since World War II, and new shipping routes opened up due to canal construction and the retreat of polar ice sheets, the risks have only grown. That around half of the non-natives established on the European Atlantic coast and in the Baltic Sea are thought to have arrived in, or on, ships, is hardly surprising.

We shouldn't forget that seagoing vessels promulgate fresh-water invaders too, including those from the Ponto-Caspian region, many of which can withstand days, if not weeks, of immersion in saltwater. Numerous terrestrial species, from flour beetles to rodents, have also stowed away on ships and their cargo. Among various land-based exotics to have colonised Britain thanks to our seafaring history is the yellow-tailed scorpion, a native of northwest Africa and southern Europe, which scampered ashore at Sheerness on Kent's Isle of Sheppey in around 1860. Some 45 millimetres long when fully grown, and in possession of a mildly venomous sting, yellow-tails still make a home amid the crumbling walls of the former dockyard. In doing so, they hold the record for the northernmost population of any scorpion in the world. At least two kinds of false widow spider – arachnids with a painful but seldom fatal bite – which are now breeding in South Devon, are also thought to have arrived with cargo in the nineteenth century. False widows have turned up in Ireland as well in the last 20 years. But it's in the marine realm that the impacts of shipping are most felt.

For much of maritime history, biofouling has been the number one pathway for the ship-borne spread of aquatic non-natives. It's easy to see why. Within minutes of being placed in seawater, any structure, including a boat, will begin accumulating bacteria, single-celled algae and other microscopic organisms. These aggregate in a slimy matrix, known as a biofilm, which prepares the ground for larger lifeforms to settle, among them seaweed, barnacles, tube worms,

mussels, clams, soft corals, bryozoans, sponges, sea squirts
and sea anemones. These fixed, or 'sessile', pioneers in turn
provide food and habitat for free-living creatures, from crabs
and shrimps to marine worms and small fish. Meanwhile,
tunnelling invertebrates will make Swiss cheese of any timber
hull. Shipworms, a specialised type of clam, are the most
notorious culprits in this respect, but gribbles, tiny aquatic
cousins of the woodlouse, also get in on the act. The galleries
and cavities produced by these creatures, dubbed 'termites of
the sea', are a refuge for still more organisms. And, of course,
hordes of parasites and pathogens also join the party.

We are aware that from the fourteenth century onwards
European vessels began moving entire marine communities
of fouling species between the northeastern Atlantic and the
shores of Africa, Asia and the Americas, but the same engine
of dispersal was doubtless whirring much, much earlier. Today,
many of the world's most troublesome marine invasives are
biofoulers. Among those yet to reach British seas is the paddle
crab from Japan and China, which passes on a lethal virus
called white-spot syndrome to farmed crustaceans. Then,
there's the Asian green mussel that is clogging the intake pipes
of harbourside factories on the eastern seaboard of the United
States and threatens to cross the Atlantic. And we shouldn't
forget the northern Pacific seastar, a fast-breeding and ravenous
predator of molluscs and fish eggs. (Not every fouling species
travels on ships: a Pacific red alga glorying in the name Captain
Pike's Golden Gate weed is believed to have arrived in the
Isles of Scilly during World War II stuck to the anchors of
American flying boats. It's now in Cornwall and could spread
around the rest of the UK.)

Biofouling is also bad news for seafarers. If a hull is left
untreated, layers many centimetres thick can form, increasing

a vessel's drag in the water and lowering manoeuvrability. The resulting loss in efficiency translates into greater fuel consumption, not to mention excess greenhouse gas emissions. Biofouling costs global shipping an estimated $260 billion every year, a figure which includes the cost of removing and preventing it. From the 1960s, tributyltin (TBT) was widely incorporated into anti-fouling paints and other coatings, but this toxic chemical was banned in 2008 after its disastrous side-effects on marine ecosystems were recognised. Other harmful treatments, such as copper-based biocides, remain in general use, but the hunt is now on for safer alternatives. These include sophisticated Teflon-like coatings that stop barnacles, mussels and other fouling organisms getting a decent footing, and chemicals – inspired by compounds which naturally occur in certain red seaweeds, corals and sponges – which inhibit the formation of biofilms. But meanwhile, an entirely new pathway for the global spread of aquatic non-natives had emerged.

As we have seen, solid ballast had long been used to stabilise merchant vessels. But loading and unloading the stuff had always been a laborious, time-consuming and expensive process. Then, during the Industrial Revolution, timber-hulled vessels started to be phased out in favour of wrought iron and steel. As well as resisting fire, shipworms and gribbles – although not biofouling – these new, stronger, materials allowed the construction of far larger vessels. They also presented ship operators with a new problem: as ever more gargantuan vessels were unleashed onto the world's oceans, relying on solid ballast was out of the question. Ship designers turned instead to water ballast which, unlike soil, rocks or chunks of iron, could be transferred to and from a vessel in precise volumes according to the weight of cargo. For a wooden ship, storing large quantities of seawater could be risky; not so for a metal one, and by the late 1880s, improvements in pumping technology were driving a wholesale shift towards water ballast, more or less completed by 1930. Although a variety of mostly terrestrial hitch-hikers had been dispersed in solid ballast over the centuries, these would prove insignificant when compared to the torrent of marine invasives that could now start travelling the oceans.

The trouble is that seawater isn't all that gets sucked into a ship's ballast tanks: innumerable aquatic lifeforms, including those from bottom sediments, get swept up too. Modern ballast intakes are fitted with strainer plates to prevent the ingress of foreign objects, but these fail to stop microbes, larvae, eggs, spores, seeds, cysts and other propagules. The grates also get damaged, allowing small fish, such as gobies, blennies and sea lampreys, to come and go. Today, seven billion tonnes of ballast water are shipped around the world every year, harbouring at any one moment upwards of 10,000 marine and coastal species, many to be released at distant locations when the tanks are discharged. It's therefore hardly surprising that ballast water transfer is blamed for some disastrous marine introductions and, in a twist of fate, one of the most catastrophic befell the Ponto-Caspian region.

The sea walnut sounds harmless enough, but when, in the 1980s, this western Atlantic variety of comb jelly was brought in ballast water to the Black Sea, havoc ensued. Comb jellies – the technical name is ctenophores (with a silent 'c') – are carnivorous and while superficially similar to jellyfish their tentacles lack stinging cells and instead trap zooplankton prey by firing natural glue at them. With few predators, a tolerance to a range of temperature and salinity conditions, and a prolific reproductive rate, sea walnuts thrived in the new environment.

Within a few years, densities of up to 500 comb jellies per cubic metre were being recorded. At the peak of the invasion in 1990, sea walnuts constituted 95 per cent of all life in the Black Sea, with a total population weighing a billion tonnes, in excess of the entire global fish landing for that year. The Black Sea's ecosystem, already degraded by fertiliser run-off and human sewage, more or less collapsed following the sea walnut invasion, with the loss of its $350-million fishery (although it's since partially recovered). During the 1990s, the invader itself declined in the Black Sea due both to the exhaustion of its food supply and the accidental introduction of a natural enemy from its native waters (another comb jelly). But, by then, the sea walnut had infested neighbouring seas, including the Azov, Marmara, Caspian and eastern Mediterranean, where the cycle threatens to repeat itself. In 2006, sea walnuts appeared in parts of the Baltic and North Seas, although, due to a preference for warmer water, don't pose an imminent danger. But ballast water hitch-hikers don't just harm industries.

Because the water is drawn from the same coastal waters into which raw sewage is often expelled, ballast tanks can incubate all manner of bacteria, viruses and other deadly microbes. In late 1991, a major cholera outbreak struck coastal Latin America – the first in the region for a century. Within a year, approximately 400,000 people from Peru to Mexico were suffering from the waterborne disease, more than 10,000 of them perishing. The origin of the epidemic remains a mystery, but the strong suspicion is that local shellfish had been contaminated by dirty ballast water from Asia, possibly the Bay of Bengal. Some dispute this, suggesting that African immigrants had brought cholera, or that ocean currents swept it to Peru during an El Niño year, when waters across the Equatorial Pacific would have been warmer than average, potentially helping to incubate the microbe. Less in doubt, though, is that vessels leaving the affected region brought cholera bacteria to the United States in their ballast tanks.

Ballast water also spreads dinoflagellates. In warm conditions, some varieties of these single-celled algae can multiply into harmful blooms called red tides that can kill sea life and sicken humans. In January 2018, a Siberian Husky dog died after eating a crab on a Suffolk shoreline; tests conducted by Cefas on the crustacean found high concentrations of a natural toxin known as Paralytic Shellfish Poison, or PSP, which some dinoflagellates produce. Again, other factors may explain this particular case, but research suggests that ballast water can spread PSP-causing dinoflagellates: in one unusual case, a single tank was found to contain 300 million of them. To date, few non-natives in British waters are definitively known to have arrived in ballast water, but there's one possible example, and it's spreading fast.

Native to the coastline and estuaries of the Yellow Sea, the Chinese mitten crab was first recorded in Europe in 1912 and made its British debut in 1935 at the Lots Road power station on the Thames. The crustacean, whose furry claws also earn it the nicknames 'Shanghai hairy crab' and 'woolhand crab', took until the 1970s to properly establish here, but has since appeared in many other major rivers including the Tyne, Humber, Ouse and Clyde. During the 1960s, the species – now present from Finland to Russia – also began popping up in North America, with DNA studies suggesting the Pacific coast population came via Europe, rather than being imported direct from Asia. So what explains the success of the Chinese mitten crab?

The main trick up its fuzzy sleeves seems to be an ability to tolerate a wide range of salinities. Indeed the crab exhibits catadromy, a specialised lifestyle which sees adults feeding in fresh or brackish waters, sometimes many hundreds of kilo-metres upriver, before returning to the sea in the autumn to breed. Such is the urge to get back downstream that they will move onto land to avoid dams and other obstacles. The males die after mating, leaving the females to overwinter in deep water, each caring for up to a million fertilised eggs. The following spring, the latter hatch into miniature larvae, called zoeae, which are released into the sea, where many doubtless get drawn into the ballast systems of passing ships.

The main complaint against the Chinese mitten crab is its habit of digging metre-long tunnels into river banks and levees, weakening them and threatening foreshore real estate with collapse. The silt released from these excavations is mean-while blamed for clogging gravels upon which many fish rely to spawn. Some populations of the crab, but not yet those in the UK, harbour dangerous parasites too, including the Oriental liver fluke, which can be fatal to humans. In their home range, the crustaceans are harvested as food – the gonads are a particular delicacy – and millions of people have been infected by the fluke. (The rivers of China aren't known for their purity, so the crab flesh also tends to be contaminated with toxic pollutants.) Despite this, gastronomic enthusiasm for the mitten crabs seems undiminished, with the crustaceans still caught and eaten wherever they occur.

Even in Britain there are occasional calls for the crustacean to be harvested; the television chef Gordon Ramsay is a fan, describing the flesh as 'much sweeter and more intense than other crab meat I've tasted'. Although apparently dug in for the long haul, the Chinese mitten crab could yet prove a fleeting visitor. Populations on both the east and west coasts of North America have fallen in recent years; in San Francisco Bay, for example, where almost a million mitten crabs were recorded between 1998 and 2000, not a single specimen has been caught since 2010. No one knows why.

Ecologists and health experts alike have long called for urgent action to regulate the management of ballast water. In fact, it's been easier to turn a supertanker than achieve international consensus. In 2017, more than 25 years after being first mooted, the International Maritime Organisation's ballast water convention at last came into force. To begin with, and where possible, the world's cargo ships are now required to exchange their ballast water 200 nautical miles (370 kilometres) from land, and in water at least 200 metres deep; the thinking is that, far out to sea, fewer organisms are present to get sucked in, and those that are expelled are less likely to survive. And, from 2024, many ships will be required to fit new systems that treat ballast water on uptake and discharge with heat, ultraviolet light, chlorine, ozone gas or in some other way to kill off any organisms. Meanwhile, the sediments which collect in the bottom of the tanks – also chock-full of undesirable stowaways – have to be disposed of carefully. But the new agreement doesn't solve all the problems.

For a start, there's always a risk that organisms flushed into mid-ocean will make their way to coastlines – although experts believe this is unlikely – and ship design typically prevents everything in the tanks from being easily removed. More troublesome is that the rules are almost impossible to police: testing ballast water takes time, a commodity in short supply in today's highly competitive shipping industry, so unscrupulous vessel operators will be tempted to cheat the system. This is the reason that major shipping nations, among them the United States and Britain, are yet to sign up.

Food wrappers, polystyrene chunks, bait pots, detached buoys, carrier bags, rope pieces, empty bottles, old shoes, knots of fishing line, tampons, absorbent wipes, drinking straws. Anyone taking a stroll on a British beach these days would sometimes be hard pushed to see anything natural amid all the trash cluttering the strandline, particularly in the aftermath of a winter storm. Given that more than 12 million tonnes of plastic litter enter the world's oceans every year – a figure predicted to triple by 2025 – it's no surprise that so much rubbish is washed or wafted onto our coasts. Public awareness of the risks that plastics pose to sealife and human health is at last growing, but less attention has been paid to what's catching a ride *on* the litter. Just as policy-makers, scientists and engineers are coming together to tackle the spread of aquatic non-natives by ships, a whole new highway of marine invasion is ramping up.

Organisms have always caught a ride on floating seaweed, logs and other natural debris, but all that plastic flowing into the sea from storm drains and sewers, or discarded by careless beach-goers and fishermen, seems to be accelerating the process. Today, the same communities of bacteria, algae and marine animals that foul hulls are now associated with plastic flotsam, and many wash up on distant shores. This is most evident after hurricanes and other extreme events: in the wake of the 2011 earthquake and tsunami in Japan, almost 300 Asian species were deposited on Hawaiian and western North American beaches.

Among recent arrivals to Britain is the Florida rocksnail, ten specimens of which were recorded clinging to fishing floats during the 2015/16 winter. The predatory whelks, which originate in warm seas from the Gulf of Mexico to the eastern Mediterranean, were found on coastlines as far apart as Cornwall and Kent. Should the rocksnails establish themselves, they could threaten native bivalves, limpets and barnacles.

Other rafting organisms recently picked up on the British coastline include tropical scallops, oysters, crabs and mussels, which had drifted for some 8,000 kilometres across the Atlantic to arrive on Dorset's Chesil Beach. More worrying is that certain toxin-producing bacteria in the genus *Vibrio* apparently have an affinity for plastic microfibres, so if these build up in the waters around Britain, that's bad news.

Despite all this, our seas appear remarkably immune to exotic invasions. At the last count, fewer than 100 non-native species were known to have naturalised in British waters, versus several thousand on land. Invertebrates, such as molluscs and crustaceans, constituted roughly two-thirds of the total, with seaweed and other algae accounting for the rest. A smattering of non-native flowering plants, fish and insects was also recorded in intermediate habitats such as beaches and estuaries. But there's no room for complacency.

Given the difficulty in sampling marine organisms, and the fact that a detailed survey of the entire British coastline has yet to be conducted, the official number is almost certainly an underestimate. After all, in San Francisco Bay alone, more than 250 introduced species were recorded during the late 1990s, representing 99 per cent of all biomass. It's telling that most of the things we know about occur close to shore. Who's to say what lurks in deeper water? (Although, until now, the types of human activities that tend to spread marine invaders – ballast water exchange, recreational watersports, offshore oil extraction, wind energy generation, aquaculture, and so on – generally occur in shallow, coastal waters.) And the fact that many marine non-natives are invertebrates or algae – organisms which, by their nature, tend to be poorly studied anyway – compounds the problem. Indeed, some of the most potentially troublesome new arrivals aren't much to look at.

For instance, a tiny new type of marine ribbon worm, native to the Pacific Ocean, which David Fenwick – of Moresk flatworm fame – recently discovered on rocks in Cornwall (and which has since turned up in Dorset), also harbours *Vibrio* bacteria related to those previously mentioned. As a result of its association with these microbes, the worm contains tetrodotoxin, a nerve agent also found in pufferfish, which is a thousand times deadlier than cyanide. No one knows how the worm got here: it could have been in ballast water or stuck to a hull. Suffice to say, we need dedicated experts like Dave to keep monitoring our coastlines.

There's no denying that the chilliness of Britain's waters has discouraged many prospective colonists. But for how much longer? With climate change set to warm our seas, an armada of eclectic new marine invasives could be just over the horizon.

II

Fighting Fire with Fire

'In the market, the natives of Jiao-zhi sell ants stored in bags
of rush mats . . . In the south, if the Gan [mandarin orange]
trees do not have this kind of ant, the fruits will all be damaged
by many harmful insects and not a single fruit will be perfect.'

Plants and Trees of the Southern Regions, Ji Han, 304 CE

'Acers, limes and nettles. They are the trinity for searches,'
said Richard Comont, gesturing at a sycamore on the far side
of the wrought-iron fence. He passed a circular tray of
stretched white cloth between the railings and, leaning over
their chin-high points (unsoftened by layers of black
Hammerite), thwacked a low-hanging branch with the
aluminium handle of his butterfly-net. Dead leaves, twigs and

other crud showered the tray, which he withdrew to inspect with the restrained fervour of a gold panner. Amid the detritus were harvestmen, aphids, tiny yellow barkflies, caterpillars, a green shieldbug. But no ladybirds.

The trinity of searches. A fitting metaphor given our environs: a narrow lane skirting the grounds of the parish church in Crediton, a small market town in mid-Devon. 'And so churchyards are quite good,' went on Richard, 'because they tend to have at least two of those three around them and aren't sprayed with as much pesticide. There's plenty of food for harlequins. Also, at this time of year, you get a lot falling off the trees with the leaves. Often land on the gravestones where they overwinter, as well as in the porches.' It was early autumn and we were in quest of an insect recently declared the UK's fastest invading species – ahead of grey squirrels, American mink, ring-necked parakeets and muntjac. Its name was the harlequin ladybird.

My companion was an entomologist who took an interest in the multicoloured beetle – a native of temperate parts of eastern Russia, Mongolia, China, the Koreas and Japan – soon after its arrival on British soil. He'd noticed a specimen in a suction-trap sample collected from Hertfordshire back in October 2003, and as the harlequins spread like wildfire, Richard and his colleagues documented their ecological impacts. Adults and larvae alike proved to be ravenous and unfussy feeders, taking not only aphids, but everything from plants, pollen and fungal spores, right up to caterpillars, and even other ladybirds. As Richard put it, 'Anything smaller than they are, they'll have a nibble and see if it's edible. If it is, they'll finish it off.' Thus, harlequins beat native counterparts to all manner of food sources, then ate the competition itself.

'We looked at biological records for ten-kilometre grid

squares across the country. When the harlequin turned up, all but one of Britain's eight commonest ladybird species either started to decline, or started to decline more rapidly,' said Richard. The two-spot ladybird was among the worst hit. 'Before 2003, two-spots were the single most abundant ladybird in Britain. They were everywhere. But within five years of the harlequin's arrival, their distribution had fallen by 44 per cent. That's extinction-level population dynamics.' Although there's no hard proof, the non-native has also been linked to a decline in other insects including the small tortoiseshell butterfly, whose nettle-eating caterpillars are thought to fall prey to marauding harlequin larvae.

As it happened, we didn't need to enter the churchyard proper to find harlequins. Further along the iron railings themselves, the insects began to appear: several glossy, button-like pupae, stuck to the underside of the cross-bar; tiny spiky black larvae scampering here and there; and the adults, whose carapaces exhibited various permutations of black, red and orange. No wonder the harlequin was named for a stock character from sixteenth-century Italian theatre noted for his many-hued patchwork costume. In fact, up to 109 different colour forms have been documented, though 80 per cent of harlequins recorded in Britain are of the 'succinea' variety, with an orange base colour, dotted with anywhere between nought and 20 black spots. A smaller proportion are classed as 'melanic', having a black base. I asked Richard how he was sure, given the diversity of the patterning, that these were harlequins and not some species of native.

'A rule of thumb is that any ladybird larger than five millimetres with brown legs will almost always be a harlequin. The only exception is a scarce ladybird which is restricted to conifers.'

That we could view all stages of life on the church railings was evidence of the harlequin's astonishing reproductive rate, a feature which helped explain its success.

'Most of our 46 native ladybirds breed once a year,' said Richard. 'Harlequins just reproduce constantly and, at outside air temperatures above 20 °C, take just three weeks from egg to adult. They've probably had two, maybe three, generations this year because of the mild spring and autumn. If it's warm enough the harlequin will keep going; I've collected viable pupae on the 16th of December.'

From the human perspective, the ladybird has proved a nuisance more than anything else. Around this time of year, as the days shorten and the weather cools, harlequins begin to congregate in sheltered locations, including inside our homes, forming aggregations many hundreds of insects strong. Elsewhere in their range, particularly in the United States, the species – there known as the 'Halloween ladybug' – is blamed for ruining fruit harvests. That's because during autumn the insects head to orchards and vineyards to stock up on sugar before overwintering and get crushed in with the produce. A single squashed ladybird can taint up to five litres of wine. This is somewhat ironic given that the harlequin has been deliberately and repeatedly introduced for more than a century to defend crops.

Farmers thought they could harness the ladybird's broad diet and insatiable appetite in their ongoing war against aphids, scale bugs and other agricultural pests. In other words, they wanted to use the harlequin ladybird as an agent of so-called 'biological control'. After an initial failed release of harlequins into Capitol Park in Sacramento, California, during the summer of 1916, ensuing decades saw numerous more successful introductions across the United States and abroad.

At first the ladybirds behaved themselves, each devouring hundreds of aphids per day while protecting crops ranging from apples and oranges to potatoes and pecans. But then one day they went rogue.

The first indication of trouble was in 1988. Swarms of Halloween ladybugs began entering homes in southeastern Louisiana and eastern Mississippi. Pale-coloured buildings were preferred, possibly for their resemblance to the limestone cliff-faces common in the beetle's native range. The species spread to other parts of the country and by 2007 had been detected in all but three of the 48 contiguous states. Despite the warning signs, harlequins continued to be exported around the world during this period. In Europe, the insect had been released in small numbers for decades: Ukraine began back in the 1960s, with larger-scale use in western Europe starting in France in 1982. By the time commercial suppliers on this side of the Atlantic belatedly removed the harlequin from their catalogues in 2004, the new American strain had passed on its invasive qualities to existing European populations. These rapidly dispersed across the continent, with the species now established in more than 20 European countries, as well as popping up as far afield as South Africa, Brazil and the Middle East. Scientists have even coined a term – 'the invasive bridgehead effect' – to describe the ladybird's unusual pattern of spread, in which a single, small population going rogue in one part of the world, is sufficient to spawn a global invasion.

Although harlequins were never officially released as biological control agents in Britain (Richard has heard tell of 'unofficial' use), that didn't prevent beetles crossing the Channel. Genetic analysis of the current harlequin population in this country shows at least five separate introductions, mostly along the south coast, suggesting individuals had flown or blown from Europe. Insects may also have arrived from Canada on imports of fruit, vegetables and flowers. With few if any natural enemies and chock-full of defensive chemicals – ladybirds taste disgusting, and harlequins especially so – the beetles swept across the country at an unprecedented rate: between 2004 and 2008, the species moved north and west at more than 100 kilometres per year, and today is the commonest ladybird in England and Wales. It is now building up numbers in eastern Scotland, although the pace of spread north of the border is far slower due to the cooler climate. What turned a well-mannered orchard sentry into a world-conquering insect menace?

One theory is that harlequins from two or more different Asian source populations may have interbred during the 1980s, spawning a new, tougher and more hostile variety. Richard reckons the captive breeding process was also to blame: large numbers of ladybirds were kept in close proximity to one another, without enough food and exposed to various pathogens in the environment. Only those insects able to resist infection and with a willingness to eat anything – including other ladybirds – survived. 'So they basically inbred and inbred until everything that held them back was wiped out of their genome,' he said. 'That little population from Louisiana went on to be invasive. You can pick up its signature anywhere in the world.'

Biological control, the deliberate release of one organism to reduce the population of another, is nothing new. For centuries, Chinese citrus growers would encourage native weaver ants to move along bamboo poles between moated trees to eat stink bugs and other insect pests, sometimes boosting the predator population with additional ant nests. The use of

domesticated cats as mouse-catchers in ancient Egypt represents a still earlier example. These days, the technique is often
divided into three broad categories – classical, inundative and
conservation control – which may be deployed individually
or all at once.

Classical control suppresses an exotic pest using one or
more organisms drawn from the pest's native range. It assumes
that the most problematic species are those that have left
predators, parasites and pathogens behind, and attempts to
solve the problem by reacquainting them with their natural
enemies. If all goes well, the introduced agents establish in
the new location.

In classical control the aim is long-term containment rather
than eradication, for should the target organism die off so
might the agent. Inundative control, by contrast, is a quicker
fix, eliminating the target through the mass release of agents,
often into confined situations such as a greenhouse, orchard
or lake. Unlike the other forms of biocontrol, this approach
is temporary, with the agents themselves dying off, necessi-
tating repeated, perhaps seasonal, introduction. For this
reason, inundative agents, which are mass-produced in
captivity, are sometimes called 'biological pesticides'.

The third kind, conservation control, recruits organisms
already present in an ecosystem to the task of tackling an
unwanted species. Recognising that predators and other
natural enemies may have been reduced, for instance, through
the overuse of pesticides, the approach seeks to boost their
populations by manipulating habitats in their favour. That
could be as simple as putting up nest boxes for insectivorous
birds or creating so-called 'beetle banks', strips of uncultivated
habitat at the margin of, or even within, arable fields, as a
refuge for predatory insects.

Until recently, biological control had been driven by financial considerations: to protect agricultural, horticultural or forest products from weeds and pests. And that made sense because, in the case of classical and inundative control, the development of effective agents is expensive and time-consuming so needs to be justified from a commercial perspective. But now its remit is being widened to encompass not just unwanted organisms that threaten profits, but those that are perceived to hurt natural ecosystems. In other words, it's been recast as a promising new weapon in the war on invasive species. Alas, fighting fire with fire can be hazardous, not least when the agent of control itself becomes the problem.

The early history of biological control is littered with disasters, often stemming from the misguided introduction of a mammal, bird, amphibian, reptile or fish. The problem is that many of these vertebrates are both opportunists and generalists, devouring not only the intended target but plenty of other species besides. Among the more infamous avian miscreants is the common myna, a member of the starling family noted for its ability to mimic human voices. The bird was introduced from India as early as 1762 to the island of Mauritius by the Count de Maudave to control red locusts. Following this apparently successful trial, the common myna enjoyed subsequent releases around the world as both biocontrol and caged pet – sometimes at the urging of acclimatisation societies – including in Australia, New Zealand, South Africa and many oceanic islands. As a result the bird is now regarded, rightly or wrongly, as one of the world's worst invasive species, accused of outcompeting or preying on endangered natives, spreading weeds, stealing crops and passing on parasites. They've tried removing common mynas from islands such as the Seychelles by shooting and trapping but such attempts have largely failed. The birds are just too smart.

In Australia, the myna was one of several animals recklessly thrown at the cane beetle, a pest of sugar plantations (and native to Australia), but the bird's negative impacts were later dwarfed by the ecological devastation caused by the cane toad. This gluttonous South American species brought to northern Queensland in 1935 hoovered up a wide range of native amphibians and invertebrates, while its poisonous glands sickened anything that tried to eat it, including snakes and crocodiles. About the only thing not to suffer from the cane toad's arrival was the cane beetle. Thankfully, some native Australian predators are learning to avoid the toad.

Plenty of mammals have also been released as biological control agents with calamitous results, the small Indian mongoose being one of the more notorious. Released by the British and others to control rodents in plantations worldwide, it attacked an array of natives, particularly ground-nesting birds. Its worst impacts have been on islands: in Fiji mongoose are blamed for wiping out the bar-winged rail; in Haiti they've killed off four types of shrew. Mongoose are also implicated in the disappearance of solenodons, rare burrowing mammals found only in Cuba and Hispaniola, as well as the decline in Japan's indigenous Amami rabbit.

Fish, too, have been introduced, usually into freshwater habitats, to suppress pests. The white amur, or Chinese grass carp, is widely used to control waterweed, as well as being a popular aquaculture species. Voracious, long-lived and fast-growing, with a record weight of 45 kilogrammes, the fish was conscripted by the Americans during the 1960s to deal with an infestation of hydrilla. Otherwise known as water-thyme, this ornamental plant had itself arrived from Sri Lanka in 1947 and was overrunning many of Florida's waterways. While grass carp certainly had a taste for hydrilla, native water plants were regarded as the *hors d'oeuvre*. And when the hydrilla was gone, a completely new waterweed moved in to take its place, a process sometimes called the 'competitor release effect'. Like other cyprinids, the grass carp is also blamed for stirring up water sediment and passing on parasites such as tapeworms. These days, with grass carp now established in eight states in the Mississippi Basin and recorded from 37 others, only neutered individuals are allowed to be released into American waterways. As with farmed Pacific oysters, that means triploid organisms, which are created by spinning the carp's chromosomes in a centrifuge. The theory is that the fish serve as an inundative rather than classical control agent, dying off once the job is done. But triploids are expensive, so batches of sterilised fish are often illegally supplemented with viable individuals, which then multiply.

In Britain, grass carp were introduced to East Anglia from Hungary as farmed stock in the 1960s, again for biocontrol of weeds; since then, the rules governing their use for this purpose in the UK have been tightened up, and grass carp today can only be stocked as an ornamental species in small, landlocked water bodies. The fish requires warmer water to reproduce so does not pose an imminent threat in this country, and given their size, coarse anglers are keen on it.

The track record of biological control with insects, mites, spiders and other invertebrates has been a happier one: as a very general rule they tend to be a bit more selective in their diet than vertebrates, but, if the story of the harlequin ladybird teaches us anything, it's that rules get broken.

One of the first uses of invertebrates in the modern era

dates to the late nineteenth century. During the 1850s, a microscopic sap-sucking bug related to aphids had been inadvertently introduced to Europe on grape vine cuttings from the United States. The bug caused phylloxera, a fatal disease that, within a decade, was devastating vineyards in both Britain and France (and remains endemic to this day). In 1873, a predatory mite was brought over from America to control the bug, but the experiment failed, and Europe's wine growers were instead only saved from calamity by grafting vines onto phylloxera-resistant rootstock from the United States. Despite this unpromising start, biological control soon chalked up a spectacular success.

In around 1868, the cottony-cushion scale insect – another cousin of the aphid with a penchant for plant sap – had arrived in San Francisco's Bay Area from Australia on shipments of ornamental acacia. Eight years later, the pests had travelled more than 600 kilometres south to orange groves outside Los Angeles, possibly hitch-hiking in a consignment of lemons. In 1888, with southern California's valuable citrus industry under threat, the US Department of Agriculture dispatched the entomologist Albert Koebele to Australia and New Zealand in search of potential natural enemies of the scale. Koebele returned the following year with various candidates, including the vedalia ladybird. When 129 of the predatory beetles were placed on an orange tree, they made short work of the scale. The vedalia was allowed to multiply and distributed to citrus growers around Los Angeles, whose fruit production tripled within the year. The beetle is still a highly effective biocontrol of cottony-cushion scale in citrus, mango and guava plantations from Brazil to Japan. Vedalia hasn't yet gone the way of the harlequin, although in 2003 a pair of specimens mysteriously turned up among ivy in the garden of a London pub despite no track record of its use in Britain, so perhaps the tale isn't quite over.

The use of herbivorous invertebrates, particularly insects, to tackle unwanted plants also has a long pedigree. A famous and somewhat convoluted example is offered by the cactus moth in Australia. The story starts in 1788 when Arthur Phillip, a British naval officer and the first governor of New South Wales, introduced prickly pear cactus to Botany Bay, having collected the plants in Brazil. Prickly pears were food for tiny cochineal scale insects, which were crushed into a crimson dye used in British soldiers' red coats. At the time, Spain monopolised the global supply of cochineal, so Phillip decided that a local source of both cactus and scale was essential to the effective running of his colony. Thanks to a friendly climate and few natural predators – apart from the cochineal insects – the plants thrived but didn't cause any trouble. Then, in the nineteenth century, a new species of prickly pear was brought, perhaps as stock fodder, and this one went berserk.

Although the authorities tried poisoning and burning the cactus, as well as crushing it with livestock-drawn rollers, the erect prickly pear, as it was known, eventually swathed almost 25 million hectares of Queensland and New South Wales in a metre-high jungle of spikes. One explanation for the success of the new cactus was an ability to exploit a landscape already disturbed by cattle grazing. Then, in 1925, an entomologist, Alan Dodd, released caterpillars of the cactus moth, a South American species known to feed on the prickly pear. The moth worked like a charm, especially in warmer parts of the country: within two years it had reduced much of the cactus to a rotting, pulpy mess, and continues to provide protection to this day. If only things had ended there.

On the strength of its Antipodean achievement, the cactus

moth was dispatched elsewhere to battle unwanted prickly pear, including Hawaii, Mauritius and South Africa. In 1957, the Commonwealth Institute of Biological Control also released the insect on the island of Nevis in the Lesser Antilles, West Indies. This proved a mistake, for the moth quickly spread across the Caribbean, reaching Florida in 1989. This breach of plant health controls means that today it threatens 79 native species of prickly pear across the United States and Mexico. Believe it or not, scientists are now seeking a biological control for the moth (a variety of wasp is in the running).

Insects have also long been used to control the scourge of St John's-wort, so called because the late June blooming of its showy yellow flowers coincides with the supposed birthday of John the Baptist. The Old World species was first introduced to North America in the 1790s, turning up in Australia and New Zealand a few decades later. Although its tiny seeds could easily have been spread to these colonies as contaminants of grain or hay, or stuck to wool-bearing animals, St John's-wort was deliberately cultivated as well, both as a garden ornamental and herbal remedy for ailments ranging from anxiety to haemorrhoids. With each plant releasing up to 33,000 seeds per season as well as reproducing vegetatively by rhizomes, the St John's-wort readily naturalised in its new range, and, like the prickly pear, probably benefited from the habitat disturbance wrought by overgrazing. Therapeutic it might have been, but for livestock St John's-wort proved toxic. Light-coloured animals – white sheep, for instance – were especially vulnerable to a red pigment within the plant that, if consumed in large quantities, left the skin painfully sensitive to sunlight. St John's-wort was dubbed 'Klamath weed' after the northern Californian river where its pest status was first recognised. It was hard to control without affecting other pasture species and resisted most known herbicides. With millions of hectares infested and land prices tumbling, desperate agriculturalists turned to biological control.

Chrysolina leaf beetles are small and herbivorous, with a lustrous, almost metallic sheen. These insects also have an incorrigible predilection for St John's-wort, and when several species collected in Europe were released into Australia in the 1930s, and later into the United States, they were successful in eradicating the weed. In California alone, livestock farmers accrued savings of more than $20 billion during the 1950s thanks to the leaf beetles. Similar successes were recorded in Canada, Chile, New Zealand and South Africa. But again, there was a minor sting in the tail: different non-indigenous plants tended to jump in to take the place of the ousted weed – another case of the competitor release effect. Worse still, the beetles sometimes got a taste for native varieties, as during the 1970s when they switched to goldwire, an American counterpart of the wort.

Despite the early promise – give or take a few hiccups – interest in biological control as a means of pest management waned during the middle of the twentieth century. The reason was simple: a cheaper, faster and seemingly more effective solution was now available to all.

In the 1940s, a new generation of synthetic chemical pesticides emerged whose efficiency and potency verged on the miraculous. Up until then, farmers, foresters, food warehouse managers and householders tended to rely on a handful of basic treatment, mostly containing lime, copper or sulphur, which weren't particularly selective and often damaged the crops and goods they were designed to protect. Then along came exciting new compounds such as aldrin, dieldrin and

2,4-D that seemingly had no such shortcomings. The most sensational of all was an odourless, colourless and tasteless powder called DDT.

First synthesised in the late nineteenth century, dichlorodiphenyltrichloroethane came into its own towards the end of World War II when the US Army applied it in several theatres of battle to successfully fight outbreaks of typhus-spreading lice and malarial mosquitoes. After the war, large quantities of DDT were sprayed against a range of insects, such as the Asian pink bollworm, a pest of cotton crops in Arizona and California. With a surfeit of planes and trained pilots in the aftermath of the war, the aerial crop-duster earned itself a place in the toolkit of the 'modern' farmer alongside artificial fertilisers and Massey Ferguson tractors. Across the world, many other weeds, insects and other undesirable organisms found themselves subjected to relentless spraying campaigns, often at the urging of powerful petrochemical corporations. (With many of the new pesticides deriving from hydrocarbons, they had much to gain.) In Britain, a parasitic wasp widely used to protect tomatoes, cucumbers, peppers and ornamental flowers from glasshouse whitefly, was one of a number of biological control agents pensioned off in favour of DDT and other organochlorine insecticides. By the mid-1950s, almost 40 different compounds were approved for agricultural use in this country, a figure rising to 136 by 1970.

Often a succession of different chemicals would be thrown at the same problem. The ongoing battle in the western USA against the Mormon cricket provides a case in point. Not true crickets, but a type of large, flightless cicada, these insects periodically coalesce into swarms of Biblical proportions, devouring crops, invading houses and even triggering states of emergency. According to legend, the insects threatened

Mormon settlers with starvation until a flock of gulls arrived to wipe them out. During the twentieth century, authorities have blasted outbreaks with any number of poisons including nicotine sulphate, sodium arsenic dust, aldrin, malathion and carbaryl. There's also America's long-running effort to control the gypsy moth, which has seen northeastern forests doused in copper acetoarsenite, lead arsenate, DDT, carbaryl, trichlorfon and diflubenzuron. Despite this cocktail of pesticides, the gypsy moth remains a pest to this day. (To be fair, biological control hasn't worked better: since 1906 more than 80 different candidates have been tested, all but a dozen failing to establish. Among the more promising had been a parasitic fly repeatedly released until the 1980s when it was found attacking native species, including an indigenous silk moth.) In the meantime, public concern began to grow about the non-target impacts of pesticides.

Just like the vertebrates deployed in biocontrol, chemical treatments were accused of harming other wildlife and generally degrading natural ecosystems. In her 1962 book *Silent Spring*, the American biologist Rachel Carson eloquently linked a decline in many birds, mammals, fish and crustaceans to the overuse of pesticides. She argued that as the toxins accumulated in food chains, top predators began to disappear. Pesticides even nullified the gains won by biological control efforts, as when DDT inadvertently killed off vedalia beetles in California but not cottony-cushion scale insects, whose numbers soared, defoliating many orchards. Humans could suffer too: DDT is now linked to cancer, male infertility, miscarriages and Alzheimer's disease, and as a result has been internationally banned since 2001 for most uses except mosquito control. But from the perspective of users there was a different problem: the pesticides stopped working.

The problem was that pests, fast breeders by definition, developed resistance. Even if 99.9 per cent of their population was killed off, the remaining 0.1 per cent soon built up numbers again. And should there be something about the survivors' behaviour or physiology that allowed them to avoid or resist the pesticide, then these attributes could be passed on and fortify the next generation. From the late 1940s, as the use of chemical pesticides ramped up (and beneficial organisms suffered), cases of resistance went through the roof. For instance, at least 28 species across 16 plant families have reportedly evolved resistance to compounds like 2,4-D, a synthetic hormone meant to control broad-leafed weeds, especially aquatic ones. Similarly, fluridone, which had once made short shrift of hydrilla, began to lose its potency in the 1990s when new resistant strains of the Asian waterweed emerged.

At least 249 types of plant worldwide are now resistant to one or more herbicides that might previously have controlled them. Among the few remaining effective treatments is glyphosate but, as we have seen, calls for its prohibition are mounting due to concerns about possible risks to human health. Animals too have developed immunity to chemical treatments, such as warfarin-resistant rats and mice. There's also a new strain of European rabbit in southwestern Australia that shrugs off a vertebrate-killing compound called 1080. But when it comes to pesticide resistance, invertebrates outstrip everything else.

The American entomologist Axel L Melander was one of the first to report the phenomenon after observing that 90 per cent of scale bugs in an orchard in Washington state withstood treatment with sulphur-lime, a hitherto effective insecticide. Writing in 1914, Melander noted that: 'Even with 26 ° sulphur-lime, ten times stronger than a normal application, 74 per cent of the scales were still alive.' Since then, almost 600 species of insect, crustacean and spider have exhibited resistance to one or more chemical treatments, with more than a dozen that no known synthetic pesticide can touch. In Britain, two types of South American leaf-mining flies that arrived in the late 1980s and attack everything from pea and potato to chrysanthemum and cucumber, are so fast-evolving that the average effective field-life of insecticides used against them is less than three years. With the shortcomings of chemical treatment becoming harder to ignore, the 1970s saw a renewed interest in alternatives to the so-called 'pesticide treadmill'. It was time for biological control 2.0.

It's not a cheap option. The mistakes of the past have led to a far more rigorous testing procedure, with years and millions of pounds invested before a new biocontrol agent is deemed safe for release. But, for classical biocontrol at least, once the successful candidate establishes itself in the wild and goes to work, no further input is required. What's more, unlike with pesticides, many biological control agents are able to co-evolve with their target species, potentially overcoming resistance. Upwards of 450 types of beetles, flies, moths and other agents, including nematode worms, have now been intentionally released against 175 weeds across 90 countries. Even more are being thrown at insects and other arthropod pests: some 2,500 agents across 196 countries or islands at the last count. Until now, the majority of introductions have occurred in Australia, North America, South Africa, Hawaii and New Zealand. This geography of use partly reflects the historical imbalance in the exchange of biodiversity between the Old World and the New, with the preponderance of pests travelling to the colonies rather than back the other way.

The remit has also widened in recent years, as concerns have mounted about the ecological, as well as economic impacts, of unwanted organisms. No longer are agents deployed merely to clobber crop pests, they're also being set on species deemed to damage natural habitats. In America, the purple loosestrife represents one of the first cases of an invasive plant targeted exclusively for environmental reasons, with a biological control programme initiated in 1985 in partnership with the International Institute of Biological Control in Switzerland. Today, four kinds of European beetles are actively suppressing purple loosestrife across the United States. Another early example was the 1993 release of a predatory ladybird – what else? – on St Helena in the southern Atlantic. The small volcanic island's population of endemic gumwood trees was on the cusp of extinction thanks to an infestation of South American scale insects (not to mention centuries of exploitation for timber and firewood, and indeed wholesale clearance in the interests of grazing livestock). Within a few years the ladybird, which had already chalked up successes in Hawaii and Africa, was doing its thing and the gumwood was saved.

Interest is growing in the deployment of pathogenic bacteria, fungi and viruses as biological control agents, given that many appear extremely host-specific. The use of myxomatosis to eradicate rabbits is still the best-known example. In 1950 myxoma, a pox virus carried harmlessly by South American forest rabbits but lethal to their European counterparts, was released in Australia. After a brief lull, the insect-borne virus started spreading at an extraordinary rate across the continent, killing off almost the entire rabbit population within three years. Myxomatosis was then deployed in France in 1952, and in the autumn of the following year the virus appeared in Britain. As elsewhere, the impact was astonishing, soon wiping out 99 per cent of all rabbits. For a few short years, as the stench of rotting rabbits wafted through the countryside, grasses and woody seedlings grew with freedom, oak forests regenerated, flowers were everywhere. But it wasn't all good.

Red kites, buzzards, foxes, stoats and other native predators for which rabbit represented a significant portion of their diet, suffered from a shortage of food. And there were less predictable effects as heathlands and grasslands, partly maintained by rabbit grazing, began to diminish. In Dorset, for instance, a kind of red ant dependent on shorter sward declined due to the lack of grazing pressure, in turn spelling extinction for the large blue, a butterfly whose larvae had parasitised the ants' nests. Populations of several other short-grass butterfly species, notably Adonis blue, chalkhill blue and silver-spotted skipper, also crashed. But with arable farmers enjoying bumper harvests, few complained. Then the rabbits bounced back.

A combination of myxomatosis losing its virulence and the evolution of genetic resistance by rabbits meant that by the 1980s, the mortality rate from the disease in the UK had fallen by 70 per cent. Although rabbit populations continue to be hit by twice-yearly outbreaks of myxomatosis, far fewer animals die, and towards the end of the twentieth century, numbers in Britain had returned to around half their pre-myxomatosis levels. In Australia, the same thing happened: an initial population crash followed by a slow recovery. But the story doesn't end there.

During the 1980s a new virus, totally unrelated to myxomatosis, but also lethal to European rabbits, emerged in China, where – depending on which report you believe – it killed between 14 and 140 million rabbits in the space of nine months. The pathogen, called rabbit haemorrhagic disease virus (RHDV), is thought to have mutated from a harmless type, known as a calicivirus, which was already endemic to the species. RHDV soon spread to around 40 countries worldwide via the trade in both live animals and their fur. In 1988, the virus reached Spain, the ancestral home of the rabbit, with devastating consequences for both the animal itself and the predators, such as lynx and imperial eagle, which relied upon it. At first, Britain's rabbits appeared resistant to RHDV: the outbreaks that started occurring from 1992 were serious but short-lived. Now, a new strain – dubbed RHDV2 – has come on the scene and seems far deadlier. In 2018, the British Trust for Ornithology reported that the UK rabbit population was 60 per cent lower than in 1995, with RHDV2 considered a major factor. As with myxomatosis, RHDV doesn't always spread by accident.

Over the last 20 years, the virus has been deliberately released in Australia and New Zealand in just the latest skirmish of their 150-year-old campaign against the rabbit. With bunny-hatred entrenched among the farming communities of both countries, further releases of more potent RHDV strains are planned. In Britain, public enthusiasm for the use of microbes in biological control is lukewarm at best. Still, that hasn't ruled out the deployment of a few.

The best-known example is Bt, which stands for *Bacillus thuringiensis*, a bacterium widely used as a so-called 'biological insecticide'. Bt, whose spores produce toxic protein crystals that rip apart the guts of caterpillars and other larval insects, is the weapon of choice against the oak processionary moth, and in other countries has been deployed against gypsy moth and Asian tiger mosquitoes. Meanwhile, in orchards a virus is now routinely used against codling moths, the ones responsible for that maggot in your apple. For more than a century, there has also been interest in fungi as possible control agents. But serious investigations only began in the 1970s, and even now just a handful of fungal pathogens have been approved in the UK. These have taken the form of inundative controls, intended for use in the confines of glasshouses, against aphids, whitefly, vine weevils and the like. Until recently, a fungus had never been released into Britain's countryside for the purposes of controlling an invasive species. That changed in 2014.

'Stinging nettles are a curse of the job,' grinned Kate Pollard. 'Luckily, I've only been stung by a bee once.' It was late summer and the Himalayan balsam stands were peaking; although a good number were still bedecked in rose-scented pink flowers, most were now weighed down with seed capsules, swollen and primed to erupt at the slightest touch. (It was a peculiarly addictive pleasure to let these horticultural firecrackers go off in your hand.) I had joined Kate at Harmondsworth Moor, close to Heathrow airport; this former gravel extraction and landfill site had been reborn in 2000 as a country park with 70,000 trees planted for wildlife and public recreation. It was also a magnet for invasive species, including a population of Himalayan balsam far more impressive than anything I had seen back in Devon. The Moor's current status was fragile, since it was in line to be tarmacked over, should Heathrow's controversial third runway ever get the go-ahead; in the meantime, it served as one of 25 balsam-infested sites across England and Wales that were hosting a pioneering experiment in biological control.

'We're looking for open areas with healthy plants. Not too

shaded,' said Kate. Wearing latex gloves, she strode towards a nearby mass of floppy stems, many of which towered over her, and choosing the first plant, worked rapidly between leaves spraying the underside of each with a light brown liquid from a small plastic bottle. She moved to the next specimen and repeated the process. It was as if she was applying pesticide, except the contents of the bottle were intended to harm the plant, not protect it. The spray's active ingredient was a rust, a variety of pathogenic fungus selected for its ability to damage Himalayan balsam. Kate told me how, two weeks after inoculating the plants, dark brown pustules, powdery to the touch, should begin appearing on the leaves, with other balsams in the area later succumbing as the dust-like spores dispersed on the wind. If all went well, the rust would also overwinter on leaves in the soil, infecting and stunting the growth of balsam seedlings the following year. Should the rust survive without help for a second winter, then the trial would be deemed a success.

Kate's work is just the latest of numerous biocontrol projects led by her employer, CABI, an international not-for-profit firm supporting agricultural and environmental managers. In recent years, the organisation has investigated the potential of several effective agents against invasive plants in Britain, including a weevil not much bigger than a full stop that is used as an inundative control to eradicate colonies of water fern, and a mite to deal with the waterborne scourge that is New Zealand pigmyweed. CABI is also evaluating a fungus for treating buddleia stumps and another weevil for floating pennywort. Of course, when it comes to invasive plants, the Holy Grail would be an effective control for Japanese knotweed. There's still hope that the Japanese psyllid bugs will do the business but, so far, the sap-suckers haven't coped well with the British winters. As far as Himalayan balsam is concerned, the signs are more encouraging, although finding the right agent to thwart its advance has been a painstaking process.

As long ago as 2006, expeditions were mounted to the plant's native range in Pakistan and India in search of promising candidates. (There is an intriguing symmetry to these modern quests, in effect the corollary to the plant-hunting mania of the past.) Early contenders included weevils and other insects that naturally attacked the balsam, but when tested in the lab, they ate other plants too, so were struck off the list. The researchers eventually settled on a rust that appeared suitably specific. Sure enough, when the fungus was screened against a series of 74 different plant species – relatives of the Himalayan balsam, as well as those likely to share its habitats – only the balsam was infected. In fact, if anything, the rust was *too* specific.

In August 2014, after six years of quarantine, and with the necessary government approvals, CABI started field trials in Berkshire, Cornwall and Middlesex using a strain of the fungus collected in India. Puzzlingly, the balsam was affected in some sites, but not others, and it turned out that more than one variety of the plant was at large in the UK.

'Himalayan balsam seems to have been introduced multiple times from different regions in its native range,' explained Kate. 'Although the plants are classed as the same species, they have slightly different genetic make-ups which means not all are attacked by the Indian rust.' Since 2017, a second strain of rust, from Pakistan, has been tried out with more promising results, and the number of trial sites has expanded. 'This new strain seems to infect a wider range of balsam populations than the Indian one,' said Kate, who nevertheless sought to manage expectations. 'This is not a quick fix, the rust will probably take five to ten years to have a measurable impact.' And, there's a further complexity: some balsams can have other types of harmless fungi and bacteria naturally present in their shoots and roots, and these microbes help the plant to fight off the rust's invasion attempts.

In the event, the agent I saw being tested did manage to survive the winter at Harmondsworth and cause significant damage to the plants the following summer, as well as at three further sites. But for many places, particularly in the southern half of England, the search continues – right now in the Kashmir region of India – for a rust strain whose effects are broad enough to cover all Britain's balsam. Meanwhile, Kate and her colleagues at CABI are at pains to stress that they are not on a mission to eradicate the species. They hope to thin out the monocultures, giving a chance for native plants such as willowherb, meadowsweet, and, yes, even stinging nettles to recover.

12

The Future

'I would feel more optimistic about a bright future for man if he spent less time proving that he can outwit Nature and more time tasting her sweetness and respecting her seniority.'

'Coon Tree', *The New Yorker*, E B White, 1956

For as long as humans have lived in Britain, plants, animals and other organisms from elsewhere in the world have been here too. They've arrived with and without our help. Most have been harmless, some useful. Every so often though, a miscreant has appeared. A fast-breeding, fast-spreading species which doesn't play by the rules. A bad actor that harms ecosystems, carries disease and interferes with our livelihoods (or our fun). But how can we tell in advance which introductions

will develop into full-fledged invaders, so we can act to miti-
gate their effects, or even better stop them arriving in the first
place?

With only a fraction of visitors ever likely to establish
themselves and, of those, only a minority misbehaving,
predicting the next grey squirrel, Japanese knotweed or signal
crayfish is something of a dark art. This requires forecasting
not just the likelihood that a new species will establish and
spread, but also that it will have negative impacts – outcomes
that aren't necessarily correlated. Charles Darwin himself
pondered these questions. One line of enquiry has sought to
identify whether particular aspects of an organism's physiology, behaviour or genetics make it invasive. Much work has
gone into compiling a checklist of characteristics signalling
that something might be a problem, a sort of invasive species
blueprint. And there are indeed a number of attributes that
many troublemakers share.

A rapid rate of reproduction is probably among the most
important invasive markers, and is evidenced in everything
from mice and muntjac to rhododendron and rabbits. But it
can be hard to disentangle cause and effect: is a high breeding
rate the secret to a successful invasion, or just the *result* of it?
Another characteristic much in evidence is the ability to reproduce asexually, offering the potential for a single unmated
individual to found an entirely new population. Many fast-
spreading plants, such as Japanese knotweed, New Zealand
pigmyweed and wakame, as well as a range of animal pests
from flatworms and stick-insects to gall wasps and comb jellies
share this trait. And we haven't even mentioned bacteria,
fungi, viruses and other microbes. But the rule has its many
awkward exceptions, not least Himalayan balsam and giant
hogweed, both of which rely on seeds for recruitment, and

it also doesn't much help us with 'higher' animals, including invasive birds and mammals, almost all of which reproduce sexually.

Could a dogged ability to endure and adapt instead be the main requisite? After all, any aspiring colonist must have a capacity to survive and flourish in unusual new habitats and, just as important, to withstand the often challenging process of getting there, be it stuck to a sheep's fleece, secreted in a ballast tank or stowed away in the cargo hold of a Boeing. Many of the world's most notorious invasives demonstrate undoubted grit, such as the zebra mussel, a freshwater mollusc which can nevertheless tolerate up to a fortnight's immersion in saltwater when the need arises. Signal crayfish are just as hardy; research in the Czech Republic found that the crustaceans shrugged off treatments of chlorinated lime at levels toxic to most other lifeforms. What's more, when crayfish-infested ponds were pumped dry in the autumn and refilled the following year, the animals reappeared as if nothing had happened even though winter temperatures had plunged to minus 20 °C. Yet, not all invaders are quite so tough. Take the European rabbit. Who could have predicted that an animal too feeble to dig its own burrows on its debut in twelfth-century Britain would become a world-conquering pest? Even today, in Spain and Portugal, rabbits remain close to extinction. So, it can't be just a simple case of survival of the fittest. Perhaps, then, we need to consider the human factor?

The story of the rabbit suggests that invasive species somehow rely on people, not only to reach a new location but to survive it too, at least in the early stages of colonisation. Weeds, sparrows, rats, cockroaches and innumerable other plants and animals undoubtedly benefit from close association with humans, and the artificial habitats we create. So, it seems possible that certain qualities predispose an organism to a life among us. The trouble is, we would be hard-pressed to identify any that aren't present in any number of species for whom humanity is probably the worst thing that ever happened to them. And, just because something has adapted to our way of life doesn't mean it will suddenly spread uncontrollably.

The blunt truth is that no single trait is common to all invaders. This was recognised as long ago as 1879 when the American botanist, Asa Gray, having tried and failed to discover 'whether weeds have any common characteristic which may give them advantage', observed that 'the reasons of predominance may be almost as diverse as the weeds themselves'. Things have moved on a bit since then, and within the plant kingdom at least, scientists are now fairly confident that certain factors, such as height, do seem to correlate with invasiveness: a 2018 review of biological traits in 1,855 varieties of native and introduced plants in temperate central Europe drew the not altogether unexpected conclusion that taller on average non-natives tended to outcompete the indigenous plants. (You only have to think of Himalayan balsam, pontic rhododendron and giant hogweed to see there could be something in this.) Nevertheless, right now, the best predictor we have that a species will become invasive in Britain is whether it has been invasive elsewhere. That's not super-helpful, because a sizeable proportion of the organisms that turn out to be a problem are first-time transgressors with no prior invasion history. Even with known offenders it's never a dead cert, because whether a new arrival flourishes or flounders apparently also depends a lot on the new surroundings in which it finds itself. Some places seem easier to invade than others.

But, again, pinpointing what makes somewhere vulnerable to invasion isn't straightforward. It's likely to include a blend of physical attributes, such as a welcoming climate, along with so-called biotic factors, such as the availability of unexploited food resources or the absence of competitors, predators and parasites. On the latter point, there's plenty to support the idea that the most troublesome invaders are those that have escaped some or all of their natural enemies – after all, that's the problem biological control is designed to fix. Introduced plants typically suffer less than natives do from disease and parasites, and they tend to be eaten by fewer herbivores. Although the pattern is not universal, the 'enemy release hypothesis', as it's known, is regarded as the best explanation for the spread in the UK of the likes of Himalayan balsam, Japanese knotweed and rhododendron, which are together attacked by fewer than a dozen varieties of insect. Similarly, it's been shown that, on average, 84 per cent fewer pathogenic fungal species and 24 per cent fewer viruses infect plants in their naturalised ranges than in their regions of origin. The same holds for many invasive animals: European starlings in America are attacked by 22 sorts of parasitic worm whereas in their home range they have to contend with 70 or more.

Meanwhile, in the seas around Britain, overexploitation of herring and sardines seems to have favoured an explosion in jellyfish numbers, which would otherwise have been contained by these predators. With a similar thing occurring in other overfished waters from the Black and Caspian Seas to the coasts of Namibia and Japan, many marine biologists now fear a global 'descent into slime'. The enemy release hypothesis does, though, have its problems.

A key challenge is to confirm that a newly colonised habitat is indeed devoid of enemies; while it's true that many recent arrivals enjoy the benefits of reduced predation and parasitism, the effect may be temporary. Research on 124 species of

non-native plants in North America found six times as many
viral and fungal pathogens associated with plants brought in
during the seventeenth century than with those introduced
just 40 years ago. Sooner or later, old enemies catch up or
new ones appear.

Another theory, that looks at the problem from the other
angle, is gaining support in recent years. It suggests that some
of the more problematic introductions, particularly predators,
benefit from the naïveté of prey that have never before encountered them. For instance, when domestic cats were introduced
to New Zealand, the native kakapos, flightless parrots now
close to extinction, seem to have relied on camouflage to
avoid the danger, but were easily detected by smell. Back in
Europe, rodents, birds and other prey that had long co-evolved
with cats, have far more effective antipredator responses. More
research is needed to understand the role of the prey naïveté
phenomenon, and how it can be used to predict invasions.

The degree to which a place is disturbed prior to the arrival
of a new species is another condition often thought to correlate
to its invasibility. A better balanced natural ecosystem is assumed
to be somehow more resistant to aggressively spreading organisms, and when that balance is disrupted by human activity
– be it agriculture, urbanisation or heavy industry – many
non-natives will prosper. Put simply, the more we mess around
with natural habitats, the more invasives we get.

Proponents of this idea blame the worldwide spread of
weeds on the ploughing up of soil and suggest that the invasions of prickly pear cactus in Australia and St John's-wort in
the United States may have been accelerated by the landscape
impacts of livestock grazing. Likewise, Japanese knotweed is
known to thrive in soils contaminated by heavy metals, and
various freshwater invaders from the Ponto-Caspian region,

including zebra mussels and certain shrimps, endure Europe's most polluted waterways. The snag is that 'the balance of nature' is a somewhat nebulous concept, more or less impossible to quantify. Furthermore, the remarkable vulnerability of island ecosystems complicates matters, with even the most pristine of archipelagos defenceless against the onslaught of introduced pigs, rats, mice, cats, goats, sheep, monkeys, snakes and malarial mosquitoes.

Whether or not an invasion occurs thus appears to depend on a complex interplay between features of the introduced organism and those of the environment. And the process is dynamic, because once a new species establishes and begins to multiply, it may itself alter the surroundings, possibly to its own advantage. (Humans aren't alone in this respect.) Such feedback loops can be extremely difficult to predict. As a result of these processes – and doubtless others yet to be determined – we can sometimes witness what I like to call an 'invasiveness switch', whereby a mild-mannered species shifts, without warning and almost overnight, into a kind of ecological monster. We've already seen these transformations in the rabbit and harlequin ladybird but another example is offered by Florida's feral chicken.

Cuban immigrants to the United States originally bred the fowl for cockfighting but let them go when the practice was outlawed during the 1970s. The Key West gypsy chickens, as they became known, proved far more aggressive than regular chooks, flying longer distances, gobbling up native lizards and invertebrates, roosting out of the reach of predators and pooping on cars. They're now regularly culled as a nuisance. The plant world also has its fair share of Jekyll and Hydes, among them the Himalayan balsam, which in the UK grows two or three times as high as it does in its native territory.

And we shouldn't forget the strange case of the narrow-leaved ragwort.

Since the nineteenth century, this South African member of the daisy family had turned up now and then in Britain as a wool alien. In its home range the species was, and still is, a scrappy, annual thing, but in Europe the ragwort unexpectedly changed into an altogether different proposition: larger, earlier flowering and a perennial. From the 1990s, this new strain began arriving from the continent – possibly as seeds stuck to muddy cars using the recently opened Channel Tunnel – and is now spreading like crazy across southeast England. Botanists say its underlying genetics are the same, but something about the new environment must have triggered the transformation. Similar switches are sometimes seen in natives too, such as rosebay willowherb; once a scarce denizen of the Scottish uplands, during the early twentieth century the species mutated into an exuberant pioneer of urban wastelands, railway embankments and roadside verges, famously rising phoenix-like from the rubble of post-Blitz London. Today, it's widespread across Britain. The cause of the shift remains a mystery, but a crossing with plants from Scandinavia or North America is suspected.

Of course, we shouldn't discount the possibility that current problem species might 'switch back' again, as aggressive aspects of their behaviour and biology are softened, or the system itself changes to adapt to or ameliorate their effects. The Canada waterweed boom and bust is a classic example. And, if invasiveness switches don't mess enough with predictions, when two or more mischief-makers start interacting with each other, all bets are off.

Although the presence of one invader might inhibit that of another, they sometimes can work together in a positive feedback loop that exacerbates negative impacts. We already saw invasional meltdown, as it's called, in a freshwater habitat with Ponto-Caspian species, but there are sundry further cases from all sorts of environments. In Australia, the explosion of the European rabbit population led to a similarly devastating invasion of red foxes, the predators having also been introduced from the Old World; while back in Britain, bunnies are thought to support healthy numbers of American mink as well as helping spread invasive plants, such as the hottentot-fig. Sheep, too, play a central role in facilitating other exotics, carrying weeds and pests in their wool and altering habitats in a way that favours other non-natives. They probably do their worst on islands, reducing native flora and opening the door to fast-growing invasive weeds, but similar effects are seen elsewhere. In Britain, sheep indirectly support the marsh frog; that's because the current landscape of the Romney Marshes, a patchwork of small fields bordered by ditches and dykes, not hedges, is maintained in the interests of sheep farming but is also vital to the survival of this exotic amphibian. Should the marshes be one day given over to arable farming, its waterways filled and hedges replaced, the frog would probably hop into oblivion.

Sunlight is sometimes said to be the best of disinfectants. Not when it comes to invasive species, it isn't. Britain's weather has frozen out many potential colonisers who prefer things a little warmer. Now anthropogenic climate change looks set to put a spanner in the works, pushing up average winter temperatures by as much as 4 °C by the end of this century. The precise impacts are hard to predict, but it's safe to say that a variety of unusual new organisms will begin to find life in and around the British Isles more welcoming. Even now, we are seeing movement in our native fauna and flora, with northwards range shifts in certain butterflies, marine molluscs, migratory birds and plants consistent with recent patterns of climate change. By the same logic, species that were once confined to more southerly parts of Europe are likely to expand their ranges to encompass the UK, with many non-natives that might once have been killed by chill now lingering. Already, anchovies, tuna, cuttlefish and other species of balmier waters are beginning to turn up in the seas around us. What's more, entirely new pathways for invasion may be created.

The retreat of polar ice caused by global warming is opening up an alternative passage for cargo ships: in 2010, just four vessels plied the 12,800-kilometre Northern Sea Route along the Arctic coast of Russia from eastern Asia to western Europe. By 2017, that figure had jumped to 300. Most vessels transport oil and gas, but Maersk Line, the world's largest shipping company, recently launched the *Venta Maersk*, a 3,600-container ice-class ship dedicated to the Vladivostok–St Petersburg run, and rival carriers are likely to follow suit. It's easy to see why: although open to maritime traffic for just four months a year, the NSR cuts east–west journey times by a third when compared with the traditional route via the Suez Canal, offering obvious benefits to shipping companies but also potentially increasing survival rates for hull-fouling and ballast species.

Changing atmospheric currents could also have an impact. At the moment, Britain is subject to prevailing southwesterly winds from the Atlantic bringing precipitation but not much else. But what happens if climate change causes those winds to swing around and blow from the continent? There's a chance that vast quantities of invertebrates, seeds and other tiny airborne propagules will be swept up from Europe and the North African landmass only to be dumped over Britain. Asian hornets, harlequin ladybirds and the western conifer-seed bug are just three troublesome pests suspected to have crossed the Channel on a favourable tailwind. All this means that there's a whole lot of new species lining up to invade.

Among them is the Indian house crow. This grey-shawled corvid is not much of a flier but that hasn't stopped it spreading from its native Asia around the world on ships. Invariably associated with humans, the crow attacks fellow birds, damages crops and hangs around on rubbish dumps posing a public health nuisance. Much to the alarm of the British authorities, it's now breeding in the Netherlands and Denmark. The sacred ibis is also waiting in the wings. For the ancient Egyptians this spectacular black-and-white wading bird with downward-curving bill represented Thoth, the god of wisdom, knowledge and writing; it also helpfully removed parasite-bearing snails from rivers. The sacred ibis was originally introduced as an ornamental from sub-Saharan Africa to France in the middle of the eighteenth century, and during the 1970s zoos began keeping it in free-living populations. The bird has since formed colonies across Europe, devouring a wealth of native species. In 2004, two ibises were enough to wipe out all 30 nests of a French colony of sandwich tern in the space of four hours. Sacred it may be but, like the Indian house crow, the ibis has a penchant for human refuse and is known to feed on slurry pits. And they say cleanliness is next to godliness . . .

It's no surprise, given their preference for warm conditions, that new amphibians and reptiles are also on the radar. Several southern European species, such as the Italian wall lizard which is already established in modest numbers in southern

England, are expected to flourish. Meanwhile, a number of habitual escapees from the pet trade may start breeding in the wild. The red-eared terrapin, also known as red-eared slider turtle, is a likely candidate. It's one of the planet's most commonly traded pet reptiles: between 1989 and 1997 an estimated 52 million were exported from North America to Europe, Asia, South Africa and Australia, mostly as hatchlings. The phenomenal success of the Teenage Mutant Ninja Turtles comics and television series fuelled the market. A hefty proportion found their way to Britain, with many dumped in the wild as they grew too large or their novelty value wore off. Concerned about the risk of invasion, in 2015 the EU banned the sale of all freshwater turtles, prompting traders to offload their stock into the wild (just like fur farmers had done with mink during the 1980s). The redundant reptiles proved remarkably robust and long-lived – 40-year-old specimens are not unheard of – and still haunt many ponds, canals and other water bodies where they feast on all manner of natives. Successful egg incubation requires sustained daily temperatures of 25 °C for two months, so red-eared terrapins don't yet reproduce in this country, but breeding does now occur in southern Europe. So who knows what might happen if and when things warm up here too.

The African clawed frog – some call it a toad – was also once a popular pet, but the chief reason for its introduction was more functional: it was a living pregnancy test. During the 1930s, the British scientist Lancelot Hogben showed that the female clawed frog would lay eggs when injected with urine from a pregnant woman. The amphibian's ovaries seemed to be stimulated by the presence of a human pregnancy hormone. This was a breakthrough because, until Hogben's discovery, rabbits or mice were being used instead, which took longer and required the dissection, and hence death, of the animal. The frogs, by contrast, were conveniently reusable. In the ensuing decades, doctors routinely sent off samples to 'frog labs', with thousands of clawed frogs used for this purpose in Britain and beyond.

By the time immunological test kits became available in the 1960s, the frog was establishing small wild colonies in South Wales and the Isle of Wight. Already a hardy species tolerating surprisingly salty water and a breadth of ambient temperatures, the clawed frog is set to benefit from warmer, wetter summers. That's a problem for our native amphibians, since the frog carries – but is immune to – chytridiomycosis; indeed, some have suggested that the trade in clawed frogs may explain the global spread of the fungus responsible for this lethal disease. (This may or may not be true; indeed, experts are still trying to unravel the chytridiomycosis story, and aren't even 100 per cent sure of the exact pathogen involved.)

An insect army may be on its way too, for which Asian hornets, harlequin ladybirds, horse-chestnut leaf miners and oak processionary moths are but outriders. Ants that form large networks of interconnected nests, known as supercolonies, are of particular concern. Alongside the 'electricity ant' we met at the start of the book, is another even more notorious species.

For centuries, the Argentine ant has been hitch-hiking from its native South America to subtropical regions worldwide, including many oceanic islands. In 2002, the insect had reportedly formed a single supercolony stretching for 6,000 kilometres along the Mediterranean coast. How and why this happens is debated. One early theory, now discounted, was that genetic bottlenecks were the cause: with just a handful of queens founding new populations, neighbouring colonies tend to be so closely related that they ignore one another and even join forces, whereas back home they would normally fight. A snag is that back in their native lands Argentine ants are also known to form supercolonies, albeit much smaller ones; so, too, do many other potentially invasive ant species that haven't yet gone global. An alternative explanation is that introduced ant populations don't suffer as much from the parasites and predators – or from other competing supercolonies – that would normally limit colony size in their home range.

Whatever the explanation, when they get going, supercolonies can spread with frightening efficiency – just a handful of soil with queens and workers is enough to start a new population – often spelling disaster for indigenous ants, with knock-on effects for other wildlife. For instance, the Californian horned lizard is declining because Argentine ants are suppressing the native ants that form its diet, while in South Africa plants such as protea flowers are suffering because the native ants upon which they rely to bury and disperse their seeds are being displaced. The Argentine ant was first recorded in Britain a century ago – there's a 1927 record from Windsor Forest – but hasn't yet been able to survive the winter. With climate change, expect that to alter in the future.

Perhaps the greatest fear from a human perspective is that exotic, pathogen-carrying insects will start breeding in Britain. Malaria, West Nile virus, yellow fever, Zika, dengue and chikungunya are just a few of the painful and sometimes deadly mosquito-borne diseases that are concerning public health experts. A 2011 survey of UK local authorities revealed a two-and-a-half-times increase in reported mosquito bites over the previous decade; warmer, wetter summers during this period are thought to have provided the perfect breeding conditions for the bloodsuckers. The Asian tiger mosquito, a daytime biter, capable of carrying at least 20 different human pathogens, is easily the most worrying.

Unlike other mosquitoes, which have more or less established everywhere they can, the tiger mosquito – a native of southeast Asia and islands across the Indo-Pacific Ocean – is a relative newcomer to Europe and still extending its range. The species is a weak flier but since the late 1970s has been spreading worldwide thanks in large part to the intercontinental trade in used vehicle tyres: the small pools of water that collect in the tyres are perfect hideaways for the larvae which are also known to stow away in shipments of lucky bamboo from China. Tiger mosquitoes are now present in Australasia, the Americas, parts of Africa and, in the last two decades, have been moving at an alarming pace across western and southern Europe. Italy currently has the worst infestation, and indeed in 2007 suffered a small outbreak of chikungunya virus as a result. The warmest and wettest parts of Britain are predicted to be hosting breeding populations of the Asian tiger mosquito as early as 2030.

New climate-driven epidemics threaten animals too, both wild and domesticated. These include bluetongue virus, a viral disease of cattle, sheep, deer and other ruminants transmitted by several types of biting midge, which causes ulcers and drooling from the mouth and nose. Once restricted to Africa and south Asia, bluetongue is now present in a dozen European countries, and at the end of 2018, bluetongue-infected cows were found in Yorkshire and Northern Ireland, having been imported from France. According to government officials, the disease doesn't pose a threat to human health or food safety, but can reduce milk yields in cattle, cause infertility in sheep and, in the worst cases, kill livestock.

The changing conditions may also nudge non-natives that are already here towards more invasive tendencies by improving survival, speeding up reproduction and worsening their impacts, while reducing the ability of natives to resist. And those species able to tolerate the changing conditions and capitalise on ever more frequent disturbances and extreme events, such as flooding, fires and droughts, will prosper at the expense of the less adaptable. At the moment many young muntjac, for instance, perish during colder weather, but the deer may begin multiplying even faster should there be a shift to warmer winters. The rise in temperatures is also predicted to accelerate the spread of water fern, water primrose, parrot's feather and other troublesome aquatic weeds. The impacts of existing pests and pathogens might worsen too as hosts become rattled by extreme weather events.

Research shows that climate change increases amphibians' vulnerability to chytridiomycosis; likewise, pests of agriculture and forestry – those already here and those on the horizon, such as the emerald ash borer – will home in on drought-stressed crops. Climate change may drive the movement of non-natives in other ways: utility companies already transfer water resources from soggy parts of Britain to drier areas, further spreading quagga and zebra mussels, demon shrimps, New Zealand pigmyweed and other aquatic invasives. More of these types of internal translocations are envisaged in the near future.

Five million holidaymakers descend on Cornwall each year. With its dazzling beaches, precipitous granite cliffs, surfable waves and quaint fishing harbours, not to mention what the tourist board calls a 'rich cultural heritage' – think pasties, tin mines and *Poldark* – it's easy to see why. I suspect, however, that few visitors stop in at the port town of Par on Cornwall's south coast. Even tour operators struggle to sell its virtues, one website conceding that the 'industrial buildings connected to the dock are somewhat incongruous' at the western end of the beach, which it nevertheless insists is 'pleasant' and 'family-friendly'. A century ago Par, like other Cornish ports, including Charlestown, Falmouth and Penzance, served as an embarkation point for china clay, copper and other valuable minerals shipped around the world. Those days are gone. It has the ambience of a place that's long past its prime. Yet, for a snapshot of the future, you could do a lot worse than take a trip to Par.

Botanical surveys show that the town has more neophytes – plants brought to Britain after 1500 – than anywhere else in Cornwall. Over the past 20 years, an astounding 225 introduced species have been recorded here: that's a third more than the county's next most neophyte-rich locations, Truro and Penzance. At one point, Par boasted Britain's second highest exotic plant diversity. What makes this place such a magnet for non-natives? I was keen to find out.

One morning in early August I met up with Phil Hunt, a retired civil engineer and expert botanist who lived in the town. Britain was still in the grip of a prolonged and record-breaking heatwave, and I feared that most self-respecting plants would have shrivelled up and died weeks ago. Such concerns were allayed though as we meandered down side-streets, my companion pointing out introductions as we went. Here was an Argentinian vervain, there a Japanese honeysuckle. Plenty of southern European plants too, mostly straggly-looking things to my untutored eye, which

Phil identified as greater quaking-grass, red valerian and tall tutsan (a relative of St John's-wort with similarly toxic berries). Peppering a patch of wasteland – the site of a fairground in Victorian times – were clumps of buddleia and knotweed. We reached the narrow Par river and followed its short course towards the sea, swallows flitting low over the water. On the banks we encountered the Duke of Argyll's teaplant. First imported from China in the eighteenth century by the aristocrat whose title it bore, this thorny shrub was actually a member of the nightshade family and now a common fixture in British hedgerows, its orange-red berries feasted upon by birds. Other non-natives proliferating in the area included Japanese rose, Chinese sea buckthorn and Himalayan cotoneaster.

As we approached the sea, I now understood what the tourist website had meant. The glistening grey-white Par sands, composed of innumerable crushed seashells and granitic china clay waste, were enough to warrant the recent development of a large static caravan park adjoining the beach, and indeed right now a family was gamely frolicking in the shallows. But this wasn't picture postcard Cornwall: large grey factory towers reared up at one end of the beach and a droning noise filled the air. 'Clay dryers,' explained Phil, referring to the dewatering process to which china clay, piped to the harbour in slurry form from a nearby quarry, was subjected prior to export. These days, only modest volumes were handled, a faint echo of Par's bustling, maritime past. Back then, many soil ballast aliens would arrive – mainly from southwestern Europe – and flourish on the port-side spoil heaps.

Par's commercial heritage could not however be responsible for the town's current exotic flora; most ship-borne introductions had long ago disappeared (although, close to the river,

Phil had showed me wallflower cabbage, a classic ballast species and probable relic from the industrial period). Instead, the mild climate seemed far more significant: a sensitivity to frost was the hallmark of many of the non-native plants here, among them the Mexican fleabane (dubbed 'the Fowey daisy'), montbretia ('one of the most successful aliens in Cornwall') and the blue passionflower. Meanwhile, amid the modest dunes backing the beach were spiky-leaved specimens of New Zealand flax, first introduced to Britain by the Treseder nurseries. There, too, could be seen spectacular cordyline shrubs, another signature Treseder introduction from New Zealand. Dubbed the 'Cornish palm' by the locals, these were perfectly adapted to Par's mild climate. Both the palm and flax were now self-seeding.

Although we didn't see any that day, dozens of other sun-lovers were thriving in Par, according to botanical records. These included the three-cornered leek – named for the triangular cross-section of its stalks; introduced from southern Europe, the leek had been a nuisance in southwest England for well over a century but was now spreading fast around the rest of the British Isles. There was also the Bermuda buttercup, a yellow-flowered species which colonised by vegetative reproduction and was now a major agricultural weed in the western Mediterranean, Canary Isles, Madeira and Australia. (Neither a buttercup nor from Bermuda, it is in fact a wood-sorrel from South Africa.) The hottentot-fig was yet another South African plant recorded in Par. Introduced to Britain as a horticultural oddity in around 1690, and later used to stabilise sand dunes, the ground-creeping succulent with pretty violet flowers got going in the wild in the late nineteenth century and is today found on cliffs across southwest England.

If further evidence was needed of Par's alien-friendly environs, Phil once found a peach tree and melon plant sprouting

on a spoil-heap close to the beach. Despite all this, he professed
to be somewhat 'agnostic' about climate change. 'I'm not sure
whether what we're seeing is just a short-term cycle or a
continual warming of the planet,' he said. Whatever the truth,
right now, when it came to exotic species, Par was a hotspot
in every sense of the word.

Par is not alone. Much of southern Britain already offers a
foretaste of what the rest of the country could look like as
the climate heats up. As we saw in Hampstead Heath and
on the Wraysbury river, urban centres, especially those in
London and the southeast, are uniquely prone to colonisation by exotic species. This is partly due to the heat-island
effect in which thermal energy, released by traffic, air condi-
tioning and other urban processes, is absorbed by buildings.
The warmth is then redistributed, elevating temperatures
and sustaining organisms that would struggle to survive in
more rural areas.

Meanwhile, the sheer volume of people and goods moving
in and out of cities via road, train, air and water, ensures that
new species are constantly arriving. By the law of averages,
sooner or later some of those are going to establish. And it's
happening at the international level too. Today's globalised
economy, driving unprecedented levels of exchange between
countries and regions, offers an unparalleled wealth of invasion
opportunities for hitch-hiking organisms. Recent research
suggests that more than a third of non-native species worldwide
have only spread to new locations since 1970 and that, for most
kinds of organism, rates of invasion are on the rise across all
continents. That means more birds, more insects, more crustaceans, more molluscs, more weeds are on their way.

Despite the many difficulties, predictions of which species

are likely to establish are getting better. The arrival of the Asian hornet in 2016 took few by surprise: experts had been monitoring its northwards progress for years. Indeed, the insect was identified in a 2014 journal article as among the top ten invasive species most likely to reach Britain. What's more, within two weeks of this article's publication, two crabs it also included were shown to have already scuttled into the country, and today only a few species from the top ten list have yet to make an appearance. (The good news is that, so far, few of those ranked 11 to 30 on the list have yet to arrive.) Building on this work, some of the same authors in 2018 identified a further 66 new plants and animals threatening to invade the European Union in the near future, of which eight were considered 'very high risk', including a Panamanian mussel, a catfish native to the Indian Ocean and the fox squirrel from North America.

At the same time, scientists have proposed indicators that predict the overall damage a new introduction might cause to an ecosystem. For instance, a metric recently developed by Jaimie Dick of Queen's University Belfast compares the feeding rate and abundance of an introduced predator with the same measures in native predators. Known as the 'relative impact potential metric' – aptly acronymised as 'RIP' – it successfully describes why organisms like topmouth gudgeon or killer shrimp have had such an impact, and can therefore be applied confidently to invaders of the future. But even if our forecasts are improving, there's often little that can be done once a new invader establishes itself. One conservationist recently told me that control of invasive species in the UK was 'woefully inadequate' – although those involved in the Non-Native Species Secretariat would beg to differ, insisting that many rapid response and eradication efforts have been successful. They also point out that Britain has been a leader across Europe in shaping strategy and policy in this area.

Nevertheless, over the last 100 years fewer than a dozen non-native species have been successfully eradicated from the wild in Britain. These include coypu and muskrat, along with two or three insect pests such as the Colorado beetle and tobacco whitefly. There's a host of reasons why control is difficult, if not futile. For a start, it costs a fortune. In 2018, it was announced that invasive rodents had been purged from South Georgia in the South Atlantic. The price tag for the four-year operation, which involved helicopter-drops of more than 300 tonnes of poisoned bait across the 170-kilometre-long island? Ten million pounds. That's more or less justifiable on a relatively small area, but such a project would be unthinkable somewhere the size of Britain. And, the costs of removing a species can rise exponentially as the target becomes scarcer, a phenomenon known as Zeno's paradox: when a population has been reduced to a few scattered individuals, locating and removing the stragglers can prove more trouble than it's worth.

This was the case with the control of ruddy duck. The North American bird, introduced to the Slimbridge Wetland Centre in Gloucestershire by the naturalist Sir Peter Scott in 1948, reproduced faster than the ducklings could be pinioned, and colonised the surrounding countryside. Among suggested reasons for their success was an extended breeding season, which allowed double-nesting, and a vacant niche for a nocturnally active, bottom-feeding, insectivorous duck. In the mid-1970s, as many as 60 breeding pairs were at large across nine counties, the population exploding to an estimated 3,400 individuals by 1991. By then the ruddy ducks were also breeding in Spain where, to the alarm of local conservationists, they began hybridising with a closely related native, the white-headed duck. This was a step too far.

Representations were made at highest levels and in 2005, with UK and EU funding, an eradication campaign was launched. By summer 2014, more than 6,500 ducks had been shot at a cost of over £5 million, leaving just ten females. Those final ducks aren't coming quietly. At the time of writing, a few holdouts are lingering in northern Britain, some even continuing to breed. An official involved in the culling programme told me that: 'There's a small number of high-risk areas that we keep an eye on and control is ongoing. The government's aim is complete eradication, but that doesn't mean you have to shoot every last one. Some birds will just die of old age.'

Ecologists are scrambling for new technologies that might cut the costs of invasive species management. In 2018, drones were used to release rat poison on the 190-hectare North Seymour Island in the Galapagos archipelago, while in the UK government officials have trialled a quadcopter fitted with infrared camera to accelerate the process of detecting Asian hornet's nests; the approach, which is premised on the fact that the colony will be a few degrees warmer than its milieu, has so far proved less effective than simple human vision. Scientists are also investigating the possibility of fitting tiny electronic radio transmitters to the larger hornets.

Growing excitement also surrounds the use of gene drives, artificially manipulated genetic codes introduced into a target species in order to spread particular traits through wild populations. Scientists have already developed gene drives that promise to render disease-carrying mosquitoes infertile – a controversial first release of engineered malarial mozzies is scheduled for the African country Burkina Faso and attempts with a dengue and Zika-spreading species have been mounted in Panama and Brazil; gene-driven rodents are also on the horizon. But given that the release of genetically modified organisms in Britain is politically problematic to say the least, the technique is unlikely to be tried out here any time soon.

Indeed, public resistance is a significant factor to be considered. Few shed a tear when Tetbury's Asian hornets were poisoned, but things are different for cuddlier creatures. Wildlife charities, such as the Royal Society for the Protection of Birds, were pilloried, and reportedly lost membership, after backing the ruddy duck eradication. Even the culling of rats to protect ground-nesting seabirds on Lundy island in the Bristol Channel during the early 2000s, and more recently on the Shiants in the Outer Hebrides, sparked protests from animal rights activists who launched letter-writing campaigns and sent funeral wreaths to the offices of the organisations involved.

At least those programmes ultimately went ahead: as we have seen, animal rights groups thwarted an effort to exterminate the grey squirrel in Italy; and plans to destroy hedgehogs in the Outer Hebrides have also been blocked amid public uproar. A similar hullabaloo would doubtless attend any proposal to eliminate ring-necked parakeets, the flamboyant residents having thoroughly charmed many a Londoner. (Although the removal of monk parakeets, a South American cousin of the ring-necked, seems to have proceeded without much fuss after a culling programme was launched by the UK government in 2011.) Britain is famously a nation of animal lovers, so it's no wonder that the many well-known organisations that right now are slaughtering muntjac deer, mink, wild boar and grey squirrels, don't tend to shout about it. And, there's already an effort to resist new rules requiring the UK's wildlife rescue centres to euthanise any 'invasive species' that they take in, including animals likely to survive. As we increasingly recognise the intelligence in animals hitherto dismissed as 'lower', as well as their capacity to sense pain, so our concern about their rights and welfare grows. Indeed, campaigns to outlaw such cruelties as boiling lobsters alive are building momentum and might one day soon result in new legal protections which could extend to all decapod crustaceans, potentially outlawing such practices as the snipping of pleopods from signal crayfish.

Removing problematic organisms can sometimes result in unintended consequences. We've already seen the competitor release effect in weed control, whereby the removal of one non-native leaves the field open for another one to invade. Unexpected consequences can happen with animal pests as well. For instance, during the 1990s a previously suppressed population of mice boomed after rats and rabbits were eradicated from Saint-Paul Island in the southern Indian Ocean. Similarly, when myxomatosis decimated rabbits on Macquarie Island, south of Australia, in the mid-1980s, the island's cat population that had been eating the bunnies switched to indigenous seabirds; the next 15 years were spent wiping out all the cats which, from 2000, saw the rabbits come back stronger than ever, devastating large areas of Macquarie's native species vegetation. The sorry story came to an end in 2014 when the final rabbit succumbed to deliberately introduced calicivirus.

Meanwhile, the ways in which control is carried out often risks harming non-target species. We've already seen this with pesticides and biological control, but the same can apply to mechanical and manual approaches. The traps set out in the 1930s to catch muskrat famously clobbered water voles, and even now, every time people wade down streams in search of Himalayan balsam, creep through woodland after muntjac or clamber over tidal rocks to get at Pacific oysters, they are probably stomping on native fauna and flora. Cases of mistaken identity also occur: Devon Wildlife Trust reports that heightened concerns over the Asian hornet have led people to unwittingly persecute the native variety. Management plans have to properly account for these unwanted side-effects.

This all means that for those most concerned about the problem of invasive species, early detection and prevention of potential miscreants should be the priority. No wonder then that the government invests most of its resources in surveillance and prevention, with teams of inspectors monitoring seaports, garden centres and other points of entry, and public awareness raised via advertising campaigns and social media. And for incursions of those non-natives deemed the most destructive, such as Asian hornets, American bullfrogs, sweet chestnut blight or killer shrimp, rapid responses are launched the moment credible reports come in. Unfortunately, the scope for preventing unwanted arrivals can be limited.

The UK has often taken a lead; for instance, in banning the sale of certain aquatic plants in 2013. But the political imperative of maintaining and boosting frictionless international trade – Brexit or no Brexit – risks trumping concerns about the unavoidable corollary of that flow of goods and people, namely, the arrival of unwanted new species. Global collaboration is critical if the issue is to be seriously addressed, yet significant change at this level can take decades; witness the 25 years spent agreeing the international ballast water convention, and even now the United States and Britain aren't yet signatories (although the UK is due to join soon, and the USA insists its rules are more stringent). Should any aspect of policy be construed to favour one nation over another, or to interfere with free trade, then expect that policy to be undermined, fought or ignored. Despite this, the Great Britain Non-Native Species Secretariat continues to do what it can to improve biosecurity measures, while developing new 'Pathway Action Plans' to address the most likely routes that unwanted new organisms might exploit.

It's hard to escape the conclusion that we're fighting a losing battle with most invasive species. But what if we don't bother to fight it at all? What if the problem has been blown out of proportion? What if they're not actually a problem at all? A controversial new school of thought has emerged which is asking these very questions. With a small industry developing around the control of invasive species, cynics contend that their supposed threat has been overblown. The botanist Ken Thompson, in his book *Where Do Camels Belong?*, notes that only a minority of introductions succeed and even fewer cause any trouble, while in *The New Wild: Why Invasive Species Will Be Nature's Salvation*, the journalist Fred Pearce celebrates the 'dynamism of alien species and the novel ecosystems they create'. Such opinions have been roundly criticised, with 'invasive species deniers' lumped in with climate change sceptics and accused of preaching a counsel of despair while ignoring the complexity of ecological invasions. Pearce, Thompson and others are also reproached for playing a numbers game: yes, only a handful of non-natives are bad actors, and few cause national-scale effects. Yet, their impacts are real and, at local scales, can indeed be severe.

There's no arguing that many of the organisms we have met cause negative social, economic and environmental impacts, but the large majority of non-native species are benign, and a few also have their positives. Rabbits, in this country at least, are now thoroughly enmeshed in natural ecosystems, supporting a range of indigenous predators from red kites to wildcats, and preserving the short-sward grasslands upon which numerous butterflies, birds and other natives rely. If the rabbits went, many of these would decline. The blossom of Himalayan balsam is still dishing out valuable nectar to honeybees and other insect pollinators long after most native flowers have shut up shop for the season (although some indigenous plants suffer reduced pollination because insects are so preoccupied with the balsam). Filter-feeders, such as quagga mussels and Pacific oysters, clean up heavily polluted aquatic environments; so do certain 'invasive' plants, notably the South American water hyacinth. Japanese kelp and wireweed off Britain's coasts sustain diverse communities of native marine life. Introduced conifers provide final refuge to the beleaguered red squirrel, while sycamores have been shown to support a greater weight of insects (mainly aphids) than any other widespread native tree of their size, not to mention the earthworms that devour their fast-decomposing leaves and the lichen and mosses coating branches in wetter climes.

And what about lobsters? Would it really be so bad if the American variety settled on this side of the Atlantic? They're pretty similar to their Old World counterparts – DNA testing is often needed to tell them apart – and, because they grow larger, they offer more flesh, making them a valuable food species. You'd also cut out all those air miles. The arrival of new organisms, whatever the reason, could be viewed as normal if not invigorating for ecosystems. If only we're a bit more patient, perhaps nature will adapt.

Meanwhile, humans continue to benefit from the cornucopia of deliberate introductions. Our diets have been immeasurably improved by non-natives; so too our leisure time, and hence our mental and physical wellbeing. Just ask the country's millions of gardeners, anglers, hunters, bird-watchers and pet owners. Entire industries depend on introduced organisms. Even for species that don't nourish, entertain or enrich, the evidence for negative impacts sometimes seems exaggerated. Beyond a handful of cases on remote oceanic islands far smaller than Britain, there are precious few open-and-shut cases of a non-native directly causing the extinction of a native. Often, other factors are at play. Often the true 'invasive species' is us.

Time and again an introduced plant or animal has been held accountable for our bad behaviour, be it grey squirrels blamed for the loss of reds in Scotland, when conifer felling has been the more significant factor, or the brown tree snakes on the western Pacific island of Guam that have taken the rap for the decline of native birds, when the overuse of DDT was the likelier culprit. If anything, non-natives tend to increase biodiversity, by providing different niches for indigenous organisms to come in and exploit, or by hybridising to create brand-new species as happened with rhododendrons. Although the crossings often prove sterile, occasionally the hybrids turn out not only to be fertile, but also significantly hardier than their parents, arguably boosting the overall resilience of the ecosystem. And we shouldn't forget that for several species whose numbers had declined in native ranges – mandarin ducks and Père David's deer, for instance – Britain has been a sanctuary, almost certainly saving them from extinction.

Many people are instinctively vexed by nature in the 'wrong place' and feel compelled to act, particularly when new arrivals are shown to displace natives, degrade ecosystems or directly harm us – even if overall biodiversity is increased. But, in the end, wherever humans go, it seems certain elements of nature will go too, with or without an invitation.

Acknowledgements

A surprisingly large number of academics, consultants, ecologists, government officers and amateur naturalists are today engaged in the study and management of invasive species; almost all were generous with their time and expertise in the creation of this work. There are an awful lot of 'Doctors' and 'Professors' out there, not always obvious; to avoid offence, I have omitted all titles in the spirit of egalitarianism.

My biggest thanks go to those who allowed me to invade their space and generally interfere with their day: Nicky Green (Bournemouth University); Adrian Brooker (City of London); James Chubb (East Devon District Council); Ali Hawkins (Exmoor National Park Authority); Kristin Waeber (Forestry Commission); Nicholas Wray (University of Bristol Botanic Garden); David Gould (University of Exeter); Daniel Mills (King's College London); Jason Hall-Spencer (Plymouth

University); Paul Kitching (Natural History Museum at Tring); Richard Comont (Bumblebee Conservation Trust); Jake Chant (Devon Wildlife Trust); Bob Hodgson, Phil Pullen, Tim Purches and Roger Smith (Devonshire Association); Mish Kennaway (Escot and the Tale Valley Trust); Christian Bensaid; Catherine Chatters and Rachel Remnant (Hampshire & Isle of Wight Wildlife Trust); Jaclyn Pearson (Isles of Scilly Seabird Recovery Project); Sarah Mason (Isles of Scilly Wildlife Trust); Rina Quinlan (Mammal Society); John Bishop and Kathryn Pack (Marine Biological Association); Dick Shaw, Carol Ellison, Kate Pollard, Corin Pratt and Nikolai Thom (CABI UK); Roger Trout (Rabbitwise-Plus); Andrew Diprose and Trevor Robinson (RootWave); Martin Harwood (Woburn Abbey Deer Farm); Norma Chapman; David Fenwick; and Phil Taylor.

A massive thank you also to the following for reading and commenting on early drafts: Mum and Dad; journalist extraordinaire Daniel Bardsley; Craig Shuttleworth (Bangor University); David Aldridge (University of Cambridge); Helen Roy (Centre for Ecology & Hydrology); Nicola Spence (Defra); Jes Søe Pedersen (University of Copenhagen); Naomi Sykes, Alan Outram and Jamie Stevens (University of Exeter); Brian Boag (James Hutton Institute); Magdalena Sorger (North Carolina State University); Jaana Jurvansuu (University of Oulu); Jaimie Dick (Queen's University Belfast); Nigel Semmence (Animal and Plant Health Agency); Paul Stebbing (Cefas); Trevor Renals (Environment Agency); Charles Smith-Jones (British Deer Society); Kevin Ackerman (Food and Environmental Ltd); Arnold Cooke; and Derek Gow. Any mistakes that remain were simply meant to be.

Much of my work rests on the work of scientists. Among those not already mentioned that I would like to thank personally for sharing findings and checking facts are: Archie Murchie

(Northern Ireland's Agri-Food and Biosciences); Mark Maltby (University of Bournemouth); Regan Early, Jen Lewis, Beth Robinson, Malene Lauritsen, Carly Ameen, Stephen Rippon, Lena Bayer-Wilfert, Jamie Cranston (University of Exeter); Will Hughes and Pamela Worrall (Hadlow College); Sarah Descamps (Universiteit Hasselt); Greger Larson (University of Oxford); Anaëlle Lemasson and James Scott (Plymouth University); and Chris Thomas (University of York).

Thanks also to all those central and local government officials who lent a hand: Niall Moore, Graham Royle, Olaf Booy, Iain Henderson, Elizabeth Gardner and Colin Pavey (Animal and Plant Health Agency); Maria Pearce (Bristol Harbour Office); Andrea Kelly (Broads Authority); David Hall (Cardiff Harbour Authority); Bryony Townhill (Cefas); Rachael Bice and Natasha Collings-Costello (Cornwall Council); Bill Horner and Tom Whitlock (Devon County Council); Chris Woodruff (East Devon AONB); Matt Brazier and Joanna Heisse (Environment Agency); Steve Carter (Forestry Commission); Susan Davidson (Marine Management Organisation); Jeremy Pyne, Jan Maclennan, David Appleton, Adrian Jowitt and Gavin Measures (Natural England); Jenny Carey-Wood (North Devon Coast AONB); Roger English and Nigel Mortimer (South Devon AONB).

I am grateful to a range of private companies, NGOs and individuals working in this area including: Charlotte Coles (AC Archaeology); Max Wade (AECOM); Emily Smith (Angling Trust); Jim Foster (Amphibian and Reptile Conservation Trust); Phil Owen (Arnolfini); Amanda Lloyd (Berkshire Mammal Group); Matt Jackson and Laura Downton (Bedfordshire, Cambridgeshire and Northamptonshire Wildlife Trust); Mark Vallance (Berkshire, Buckinghamshire and Oxfordshire Wildlife Trust); Neil Green and Jen Nightingale

(Bristol Zoological Society); Ian Danby (British Association for Shooting and Conservation); Robin Marshall-Ball; Richard Vaux; Jeremy Early (Bees, Wasps and Ants Recording Society); Kevin Walker (Botanical Society of Britain and Ireland); Diane Roberts (British Beekeepers Association); Norbert Maczey (CABI UK); Sam Bridgewater (Clinton Devon Estates); Sally Curzon (Gloucestershire Beekeepers); Camilla Morrison-Bell (British Ecological Society); Simon Forrester (British Pest Control Association); David Noble (British Trust for Ornithology); Margaret Palmer and Andrew Whitehouse (Buglife); Carl Adey (Complete Weed Control); John Morris (Chiltern Woodlands Project); Kevin Claxton (Council for British Archaeology); Bob Ring (Crayaway); Ed Parr Ferris and Izzy Moser (Devon Wildlife Trust); Mark Peasley (Libraries Unlimited); Dean Woodfin Jones (Lundy Company); Chrissy Mason (Moor Than Meets the Eye); Jenny Hawley and Beth Halski (Plantlife); Tom Cadbury (Royal Albert Memorial Museum, Exeter); John David (Royal Horticultural Society); Thomas Churchyard and Paul St Pierre (Royal Society for the Protection of Birds); Peter Walker (RSK Group plc); Stuart Tyler (South West Heritage Trust); Mary-Jane Attwood (Vincent Wildlife Trust); Joe Pecorelli (Zoological Society of London); Maria Thereza Alves; Jill Beagley; Colin French; Martyn Hocking; Phil Hunt; Hazel Jones; Malcolm Lee; Mervyn Newman; and Mark Spencer.

Encouragement in the early stages from two of Britain's finest nature writers, Patrick Barkham and Stephen Moss, was warmly welcomed. Each has churned out several masterpieces in the time it's taken me to cobble together this effort. And, not forgetting the great Sir Christopher Lever (OK, one title, then), among the first to mine this rarefied literary seam.

Thank you to Myles Archibald at HarperCollins for helping

me dream up this idea and his colleagues Hazel Eriksson for finishing touches of perfection and Helen Ellis for splendid work on publicity.

Lastly, I couldn't have done it without the infinite patience, support and enthusiasm of Clair, Merryn and Hannah.

Further Reading

The science of invasive non-native species is bewildering and burgeoning, with hundreds of academic papers published annually and a growing number of dedicated journals, including *Aliens, Aquatic Invasions, BioControl, Biological Control, Biological Invasions, Invasive Plant Science and Management* and *NeoBiota*. A fantastic source of reliable information on problematic organisms is available on the websites of CABI (www.cabi.org) and the GB Non-Native Species Secretariat (www.nonnativespecies.org).

And for those interested in (other) books on the subject:

Nicholas Ashton. 2017. *Early Humans* (Collins New Naturalist Library, Book 134). London: HarperCollins.

Yvonne Baskin. 2002. *A Plague of Rats and Rubbervines: The Growing Threat of Species Invasions.* Washington DC: Island Press.

Olaf Booy et al. 2015. *Field Guide to Invasive Plants and Animals in Britain.* London: Bloomsbury.

Chris Bright. 1998. *Life Out of Bounds.* London: Earthscan.

Helen Bynum & William Bynum. 2014. *Remarkable Plants That Shape Our World.* London: Thames & Hudson.

Marc W Cadotte et al. 2006. *Conceptual Ecology and Invasion Biology: Reciprocal Approaches to Nature*. Dordrecht: Springer.

Mick N Clout & Peter A Williams. 2009. *Invasive Species Management: A Handbook of Principles and Techniques* (Techniques in Ecology and Conservation). Oxford: Oxford University Press.

George W Cox. 2004. *Alien Species and Evolution: The Evolutionary Ecology of Exotic Plants, Animals, Microbes, and Interacting Native Species*. Washington DC: Island Press.

Bob Devine. 1998. *Alien Invasion: America's Battle with Non-Native Animals and Plants*. Washington DC: National Geographic Society.

J A Drake et al. (Editors). 1989. *Biological Invasions: A Global Perspective*. London: John Wiley & Sons.

H L Edlin (Foreword by the Earl of Radnor). 1961. *New Forest Forestry Commission Guide*. Bristol: Forestry Commission.

Robert A Francis (Editor). 2012. *A Handbook of Global Freshwater Invasive Species*. London: Routledge.

Oliver L Gilbert. 1991. *The Ecology of Urban Habitats*. London: Chapman & Hall.

Cang Hui & David M Richardson. 2017. *Invasion Dynamics*. Oxford: Oxford University Press.

M A Huston. 1994. *Biological Diversity: The Coexistence of Species on Changing Landscapes*. Cambridge: Cambridge University Press.

Jackson Landers. 2012. *Eating Aliens: One Man's Adventures Hunting Invasive Animal Species*. North Adams, MA: Storey Publishing.

Christopher Lever. 2009. *The Naturalized Animals of Britain and Ireland*. London: New Holland.

Julie L Lockwood et al. 2007. *Invasion Ecology*. London: Blackwell Publishing.

Tim Low. 2002. *Feral Future: The Untold Story of Australia's Exotic Invaders*. Chicago; London: University of Chicago Press.

Richard Mabey. 1996. *Flora Britannica*. London: Sinclair-Stevenson.

Suellen May. 2006. *Invasive Aquatic and Wetland Animals* (Invasive Species). New York, NY: Chelsea House.

Suellen May. 2006. *Invasive Terrestrial Animals* (Invasive Species). New York, NY: Chelsea House.

Suellen May. 2007. *Invasive Microbes* (Invasive Species). New York, NY: Chelsea House.

Jeffrey A McNeely. 2001. *The Great Reshuffling: Human Dimensions of Invasive Alien Species*. Cambridge: IUCN.

George Monbiot. 2013. *Feral: Searching for Enchantment on the Frontiers of Rewilding*. London: Allen Lane.

Harold A Mooney & Richard J Hobbs (Editors). 2000. *Invasive Species in a Changing World*. Washington DC: Island Press.

Kelsi Nagy & Phillip David Johnson. 2013. *Trash Animals: How We Live with Nature's Filthy, Feral, Invasive, and Unwanted Species*. Minneapolis, MN: University of Minnesota Press.

T P O'Connor & N J Sykes (Editors). 2010. *Extinctions and Invasions: A Social History of British Fauna*. Oxford: Windgather Press.

Tao Orion. 2015. *Beyond the War on Invasive Species. A Permaculture Approach to Ecosystem Restoration*. White River Junction, VT: Chelsea Green Publishing Co.

Fred Pearce. 2015. *The New Wild: Why Invasive Species Will Be Nature's Salvation*. London: Icon Books.

Theo Pike. 2014. *Pocket Guide to Balsam Bashing: And How to Tackle Other Invasive Non-native Species*. Ludlow: Merlin Unwin.

David Pimentel. 2002. *Biological Invasions: Economic and Environmental Costs of Alien Plant, Animal, and Microbe Species*. Boca Raton, FL; London: CRC.

Alice Roberts. 2017. *Tamed: Ten Species That Changed Our World*. London: Hutchinson.

Andrew Robinson et al. 2017. *Invasive Species: Risk Assessment and Management.* Cambridge: Cambridge University Press.

Ian D Rotherham & Robert A Lambert (Editors). 2011. *Invasive and Introduced Plants and Animals: Human Perceptions, Attitudes and Approaches to Management.* London: Earthscan.

Gregory M Ruiz & James T Carlton. 2003. *Invasive Species: Vectors and Management Strategies.* Washington DC: Island Press.

Dov F Sax et al. 2005. *Species Invasions: Insights into Ecology, Evolution, and Biogeography.* Sunderland, MA: Sinauer Associates.

Nanako Shigesada & Kohkichi Kawasaki. 1997. *Biological Invasions: Theory and Practice.* Oxford: Oxford University Press.

Daniel Simberloff. 2013. *Invasive Species: What Everyone Needs to Know.* New York, NY: Oxford University Press.

Clive A Stace & Michael J Crawley. 2015. *Alien Plants* (Collins New Naturalist Library, Book 129). London: HarperCollins.

Uwe Starfinger (Editor). 1998. *Plant Invasions: Ecological Mechanisms and Human Responses.* Leiden: Backhuys.

Kim Todd. 2001. *Tinkering with Eden: A Natural History of Exotics in America.* New York, NY: W W Norton & Co.

Colin Tudge. 1999. *Neanderthals, Bandits and Farmers: How Agriculture Really Began.* New Haven, CT: Yale University Press.

Gill Williams. 2011. *100 Alien Invaders: Animals and Plants that are Changing our World.* London: Bradt Travel Guides (Wildlife Guides).

Mark Williamson. 1996. *Biological Invasions.* London: Chapman & Hall.

R Wittenberg & M J W Cock (Editors). 2001. *Invasive Alien Species: A Toolkit of Best Prevention and Management Practices.* Wallingford, Oxon: CAB International.

Index of Species

Species are listed by common name, but Latin names are given for clarity.

General Index